# Lecture Notes
# in Business Information Processing   174

Series Editors

Wil van der Aalst
  *Eindhoven Technical University, The Netherlands*
John Mylopoulos
  *University of Trento, Italy*
Michael Rosemann
  *Queensland University of Technology, Brisbane, Qld, Australia*
Michael J. Shaw
  *University of Illinois, Urbana-Champaign, IL, USA*
Clemens Szyperski
  *Microsoft Research, Redmond, WA, USA*

David Aveiro
José Tribolet
Duarte Gouveia (Eds.)

# Advances in Enterprise Engineering VIII

4th Enterprise Engineering
Working Conference, EEWC 2014
Funchal, Madeira Island, Portugal, May 5-8, 2014
Proceedings

 Springer

Volume Editors

David Aveiro
University of Madeira, Funchal, Portugal
E-mail: daveiro@uma.pt

José Tribolet
INESC and University of Lisbon, Portugal
E-mail: jose.tribolet@ist.utl.pt

Duarte Gouveia
University of Madeira, Funchal, Portugal
E-mail: duarte.gouveia@gmail.com

ISSN 1865-1348　　　　　　　　　　　e-ISSN 1865-1356
ISBN 978-3-319-06504-5　　　　　　　e-ISBN 978-3-319-06505-2
DOI 10.1007/978-3-319-06505-2
Springer Cham Heidelberg New York Dordrecht London

Library of Congress Control Number: 2014936308

© Springer International Publishing Switzerland 2014
This work is subject to copyright. All rights are reserved by the Publisher, whether the whole or part of
the material is concerned, specifically the rights of translation, reprinting, reuse of illustrations, recitation,
broadcasting, reproduction on microfilms or in any other physical way, and transmission or information
storage and retrieval, electronic adaptation, computer software, or by similar or dissimilar methodology
now known or hereafter developed. Exempted from this legal reservation are brief excerpts in connection
with reviews or scholarly analysis or material supplied specifically for the purpose of being entered and
executed on a computer system, for exclusive use by the purchaser of the work. Duplication of this publication
or parts thereof is permitted only under the provisions of the Copyright Law of the Publisher's location,
in ist current version, and permission for use must always be obtained from Springer. Permissions for use
may be obtained through RightsLink at the Copyright Clearance Center. Violations are liable to prosecution
under the respective Copyright Law.
The use of general descriptive names, registered names, trademarks, service marks, etc. in this publication
does not imply, even in the absence of a specific statement, that such names are exempt from the relevant
protective laws and regulations and therefore free for general use.
While the advice and information in this book are believed to be true and accurate at the date of publication,
neither the authors nor the editors nor the publisher can accept any legal responsibility for any errors or
omissions that may be made. The publisher makes no warranty, express or implied, with respect to the
material contained herein.

*Typesetting:* Camera-ready by author, data conversion by Scientific Publishing Services, Chennai, India

Printed on acid-free paper

Springer is part of Springer Science+Business Media (www.springer.com)

# Preface

Enterprise engineering is an emerging discipline that studies enterprises from an engineering perspective. It means that enterprises are studied as being purposely designed and implemented systems. Enterprise engineering is rooted in both the organizational sciences and the information system sciences. The rigorous integration of these traditionally disjoint scientific areas has become possible after the recognition that communication is a form of action. The operating principle of organizations is that actors enter into and comply with commitments, and in doing so bring about the business services of the enterprise. This important insight clarifies the view that that enterprises belong to the category of social systems, i.e., its active elements (actors) are social individuals (human beings). The unifying role of human beings makes it possible to address problems in a holistic way, to achieve unity and integration in bringing about any organizational change.

Also when regarding the implementation of organizations by means of modern information technology (IT), enterprise engineering offers innovative ideas. In a similar way as the ontological model of an organization is based on atomic elements (namely, communicative acts), there is an ontological model for IT applications. Such a model is based on a small set of atomic elements, such as data elements and action elements. By constructing software in this way, the combinatorial effects (i.e., the increasing effort it takes in the course of time to bring about a particular change) in software engineering can be avoided.

The development of enterprise engineering requires the active involvement of a variety of research institutes and a tight collaboration between them. This is achieved by a continuously expanding network of universities and other institutes, called the CIAO! Network (www.ciaonetwork.org). Since 2005 this network has organized the annual CIAO! Workshop, and since 2008 its proceedings have been published as *Advances in Enterprise Engineering* in the Springer LNBIP series. From 2011 on, this workshop was replaced by the Enterprise Engineering Working Conference (EEWC). This book contains the proceedings of the fourth EEWC, which was held in Funchal, Madeira Island, Portugal.

May 2014

David Aveiro
José Tribolet
Duarte Gouveia

# Enterprise Engineering – The Manifesto

## Introduction

This manifesto presents the focal topics and objectives of the emerging discipline of enterprise engineering, as it is currently theorized and developed within the CIAO! Network. There is close cooperation between the CIAO! Network (www.ciaonetwork.org) and the Enterprise Engineering Institute (www.ee-institute.com) for promoting the practical application of enterprise engineering. The manifesto comprises seven postulates, which collectively constitute the *enterprise engineering paradigm* (EEP).

## Motivation

The vast majority of strategic initiatives fail, meaning that enterprises are unable to gain success from their strategy. Abundant research indicates that the key reason for strategic failures is the lack of coherence and consistency among the various components of an enterprise. At the same time, the need to operate as a unified and integrated whole is becoming increasingly important. These challenges are dominantly addressed from a functional or managerial perspective, as advocated by management and organization science. Such knowledge is necessary and sufficient for managing an enterprise, but it is inadequate for bringing about changes. To do that, one needs to take a constructional or engineering perspective. Both organizations and software systems are complex and prone to entropy. This means that in the course of time, the costs of bringing about similar changes increase in a way that is known as combinatorial explosion. Regarding (automated) information systems, this has been demonstrated; regarding organizations, it is still conjecture. Entropy can be reduced and managed effectively through modular design based on atomic elements. The people in an enterprise are collectively responsible for the operation (including management) of the enterprise. In addition, they are collectively responsible for the evolution of the enterprise (adapting to needs for change). These responsibilities can only be borne if one has appropriate knowledge of the enterprise.

## Mission

Addressing the challenges mentioned above requires a paradigm shift. It is the mission of the discipline of enterprise engineering to develop new, appropriate theories, models, methods and other artifacts for the analysis, design, implementation, and governance of enterprises by combining (relevant parts of) management and organization science, information systems science, and computer science. The ambition is to address (all) traditional topics in said disciplines from the enterprise engineering paradigm. The result of our efforts should be theoretically rigorous and practically relevant.

# Postulates

### Postulate 1

In order to perform optimally and to implement changes successfully, enterprises must operate as a unified and integrated whole. *Unity* and *integration* can only be achieved through *deliberate enterprise development* (comprising design, engineering, and implementation) and *governance*.

### Postulate 2

Enterprises are essentially social systems, of which the elements are human beings in their role of *social individuals*, bestowed with appropriate *authority* and bearing the corresponding *responsibility*. The *operating principle* of enterprises is that these human beings enter into and comply with *commitments* regarding the products (services) that they create (deliver). Commitments are the results of *coordination acts*, which occur in universal patterns, called *transactions*.

Note. Human beings may be supported by technical artifacts of all kinds, notably by ICT systems. Therefore, enterprises are often referred to as socio-technical systems. However, only human beings are responsible and accountable for what the supporting technical artifacts do.

### Postulate 3

There are two distinct perspectives on enterprises (as on all systems): *function* and *construction*. All other perspectives are a subdivision of one of these. Accordingly, there are two distinct kinds of models: *black-box models* and *white-box models*. White-box models are *objective*; they regard the construction of a system. Black-box models are *subjective*; they regard a function of a system. *Function is not a system property* but a relationship between the system and some stakeholder(s). Both perspectives are needed for developing enterprises.

Note. For convenience sake, we talk about the business of an enterprise when taking the function perspective of the customer, and about its *organization* when taking the construction perspective.

### Postulate 4

In order to manage the complexity of a system (and to reduce and manage its entropy), one must start the constructional design of the system with its *ontological model*. This is a fully implementation-independent model of the *construction* and the *operation* of the system. Moreover, an ontological model has a *modular* structure and its elements are (ontologically) *atomic*. For enterprises the metamodel of such models is called *enterprise ontology*. For information systems the meta model is called *information system ontology*.

Note. At any moment in the lifetime of a system, there is only one ontological model, capturing its actual construction, though abstracted from its implementation. The ontological model of a system is comprehensive and concise, and extremely stable.

**Postulate 5**

It is an *ethical necessity* for bestowing authorities on the people in an enterprise, and having them bear the corresponding responsibility, that these people are able to *internalize* the (relevant parts of the) *ontological model* of the enterprise, and to constantly validate the correspondence of the model with the operational reality.

Note. It is a duty of enterprise engineers to provide the means to the people in an enterprise to internalize its ontological model.

**Postulate 6**

To ensure that an enterprise operates in compliance with its *strategic concerns*, these concerns must be transformed into generic functional and constructional *normative principles*, which guide the (re-)development of the enterprise, in addition to the applicable specific requirements. A coherent, consistent, and hierarchically ordered set of such principles for a particular class of systems is called an *architecture*. The collective architectures of an enterprise are called its *enterprise architecture*.

Note. The term "architecture" is often used (also) for a model that is the outcome of a design process, during which some architecture is applied. We do not recommend this homonymous use of the word.

**Postulate 7**

For achieving and maintaining unity and integration in the (re-)development and operation of an enterprise, organizational measures are needed, collectively called *governance*. The *organizational competence* to take and apply these measures on a continuous basis is called *enterprise governance*.

May 2014                                                                                    Jan L.G. Dietz

# Organization

EEWC 2014 was the fourth Working Conference resulting from a series of successful CIAO! Workshops over the years, the EEWC 2011, the EEWC 2012, and the EEWC 2013. These events were aimed at addressing the challenges that modern and complex enterprises are facing in a rapidly changing world. The participants in these events share the belief that dealing with these challenges requires rigorous and scientific solutions, focusing on the design and engineering of enterprises.

This conviction led to the idea of annually organizing an international working conference on the topic of enterprise engineering, in order to bring together all stakeholders interested in making enterprise engineering a reality. This means that not only scientists are invited, but also practitioners. Next, it also means that the conference is aimed at active participation, discussion, and exchange of ideas in order to stimulate future cooperation among the participants. This makes EEWC a working conference contributing to the further development of enterprise engineering as a mature discipline.

The organization of EEWC 2014 and the peer review of the contributions to EEWC 2014 were accomplished by an outstanding international team of experts in the fields of enterprise engineering. The following is the organizational structure of EEWC 2014.

## Advisory Board

| | |
|---|---|
| Jan L.G. Dietz | Delft University of Technology, The Netherlands |
| Antonia Albani | University of St. Gallen, Switzerland |

## General Chair

| | |
|---|---|
| José Tribolet | INESC and University of Lisbon, Portugal |

## Program Chair

| | |
|---|---|
| David Aveiro | University of Madeira, Madeira Interactive Technologies Institute and Center for Organizational Design and Engineering - INESC INOV Lisbon, Portugal |

## Organizing Chairs

David Aveiro  University of Madeira, Madeira Interactive Technologies Institute and Center for Organizational Design and Engineering - INESC INOV Lisbon, Portugal
Duarte Gouveia  University of Madeira, Portugal

## Program Committee

Alberto Silva  INESC and University of Lisbon, Portugal
Antonia Albani  University of St. Gallen, Switzerland
Artur Caetano  University of Lisbon, Portugal
Bernhard Bauer  University of Augsburg, Germany
Birgit Hofreiter  Vienna University of Technology, Austria
Carlos Pascoa  Portuguese Air Force Academy, Portugal
Christian Huemer  Vienna University of Technology, Austria
Duarte Gouveia  University of Madeira, Portugal
Eduard Babkin  Higher School of Economics, Nizhny Novgorod, Russia
Emmanuel Hostria  Rockwell Automation, USA
Eric Dubois  Public Research Centre - Henri Tudor, Luxembourg
Erik Proper  Public Research Centre - Henri Tudor, Luxembourg
Florian Matthes  Technical University of Munich, Germany
Geert Poels  University of Ghent, Belgium
Gil Regev  École Polytechnique Fédérale de Lausanne, Switzerland
Graham Mcleod  University of Cape Town, South Africa
Hans Mulder  University of Antwerp, Belgium
Jan Dietz  Delft University of Technology, The Netherlands
Jan Hoogervorst  Sogeti Netherlands, The Netherlands
Jan Verelst  University of Antwerp, Belgium
João Pombinho  University of Lisbon, Portugal
Joaquim Filipe  School of Technology of Setúbal, Portugal
Joop de Jong  Mprise, The Netherlands
Jorge Sanz  IBM Research at Almaden, California, USA
Joseph Barjis  Czech Technical University in Prague, Czech Republic
Junichi Iijima  Tokyo Institute of Technology, Japan
Khaled Gaaloul  Public Research Centre - Henri Tudor, Luxembourg
Linda Terlouw  ICRIS, The Netherlands
Marcello Bax  Federal University of Minas Gerais, Brazil

| | |
|---|---|
| Martin Op 'T Land | Capgemini, The Netherlands; University of Antwerp, Belgium |
| Mauricio Almeida | Federal University of Minas Gerais, Brazil |
| Miguel Mira Da Silva | University of Lisbon, Portugal |
| Natalia Aseeva | Higher School of Economics, Nizhny Novgorod, Russia |
| Niek Pluijmert | INQA Quality Consultants, The Netherlands |
| Nuno Castela | Polytechnic Institute of Castelo Branco, Portugal |
| Olga Oshmarina | Higher School of Economics, Nizhny Novgorod, Russia |
| Paul Johannesson | Stockholm University, Sweden |
| Pedro Sousa | INESC and University of Lisbon, Portugal |
| Peter Loos | University of Saarland, Germany |
| Philip Huysmans | University of Antwerp, Belgium |
| Renata Baracho | Federal University of Minas Gerais, Brazil |
| Robert Lagerström | KTH - Royal Institute of Technology, Sweden |
| Robert Pergl | Czech Technical University in Prague, Czech Republic |
| Rony Flatscher | Vienna University of Economics and Business Administration, Austria |
| Sérgio Guerreiro | Lusófona University, Lisbon, Portugal |
| Sanetake Nagayoshi | Waseda University, Japan |
| Steven Van Kervel | Formetis, The Netherlands |
| Stijn Hoppenbrouwers | Radboud University Nijmegen, The Netherlands |
| Sybren De Kinderen | Public Research Centre - Henri Tudor, Luxembourg |
| Ulrich Frank | University of Duisburg-Essen, Germany |
| Wolfgang Molnar | Public Research Centre - Henri Tudor, Luxembourg |

# Table of Contents

## On Enterprise Engineering and DEMO

The Nature of the Enterprise Engineering Discipline .................. 1
    *Marne de Vries, Aurona Gerber, and Alta van der Merwe*

What Does DEMO Do? A Qualitative Analysis about DEMO in
Practice: Founders, Modellers and Beneficiaries ...................... 16
    *Céline Décosse, Wolfgang A. Molnar, and Henderik A. Proper*

A Pension System Redesign Case - Limitations and Improvements on
DEMO .............................................................. 31
    *Akiyoshi Araki and Junichi Iijima*

A New Action Rule Syntax for DEmo MOdels Based Automatic
worKflow procEss geneRation (DEMOBAKER) ...................... 46
    *Carlos Figueira and David Aveiro*

## On Value and Finance

Detailed Analysis of REA Ontology ................................ 61
    *Frantisek Hunka and Jaroslav Zacek*

Evaluating Accounting Information Systems That Support Multiple
GAAP Reporting Using Normalized Systems Theory ................. 76
    *Els Vanhoof, Philip Huysmans, Walter Aerts, and Jan Verelst*

Modeling Financial Statement Preparation of a SME Enterprise by an
Accountancy Firm ................................................. 91
    *Joop de Jong*

Linking Value Chains – Combining e3Value and DEMO for Specifying
Value Networks .................................................... 105
    *João Pombinho, José Tribolet, and David Aveiro*

## On Business Processes and Use Cases

ECO-FOOTPRINT: An Innovation in Enterprise System Customization
Processing ........................................................ 120
    *Yun Wan and Vishnupriya Kalidindi*

Automatic Model Transformation for Enterprise Simulation ........... 136
    *Yang Liu and Junichi Iijima*

Introducing a Framework for Scalable Dynamic Process Discovery ...... 151
   *David Redlich, Wasif Gilani, Thomas Molka, Marc Drobek,*
   *Awais Rashid, and Gordon Blair*

From Business Process Models to Use Case Models: A Systematic
Approach ................................................................ 167
   *Estrela Ferreira Cruz, Ricardo J. Machado, and*
   *Maribel Yasmina Santos*

Approach for Semi-automatic Extraction of Business Vocabularies and
Rules from Use Case Diagrams ..................................... 182
   *Tomas Skersys, Paulius Danenas, and Rimantas Butleris*

**Author Index** ................................................................ 197

# The Nature of the Enterprise Engineering Discipline

Marne de Vries[1], Aurona Gerber[2], and Alta van der Merwe[3]

[1] Department of Industrial and Systems Engineering,
University of Pretoria, Pretoria, South Africa
`marne.devries@up.ac.za`
[2] Centre for Artificial Intelligence Research, CSIR Meraka as well as
Department of Informatics, University of Pretoria, Pretoria, South Africa
`agerber@csir.co.za`
[3] Department of Informatics, University of Pretoria, Pretoria, South Africa
`alta.vdm@up.ac.za`

**Abstract.** Enterprise engineering originated as a practice with most publications focusing on the practical facets without the underlying scientific foundation. Foundational works emerged from different authors in recent years, including Dietz, Hoogervorst and Giachetti. According to Gregor, the bodies of knowledge or theories encompassed in a discipline need to address *questions related to four classes* namely: the domain, structural or ontological, epistemological, and socio-political. As a departure point for setting a research agenda for EE, we argue that the *four classes of questions* could also serve as a basis to determine an EE research agenda. In this paper we argue that a research agenda for EE should start with the *first class of questions*, concerning the domain of the discipline and suggest that an existing model, the Enterprise Evolution Contextualisation Model (EECM), could be used to define the domain of the EE discipline.

**Keywords:** Enterprise engineering, enterprise engineering discipline, enterprise engineering research agenda, enterprise engineering theories.

## 1 Introduction

Enterprises, a phenomena that has been in existence for centuries, originate when man and machine are organised to pursue some common goal [1]. Early research on enterprise performance and improvement, emphasised the mechanistic nature of the enterprise, neglecting its systemic and social characteristics [1; 2]. As both society and technology increase in complexity, so does the enterprise. It is increasingly problematic to understand, design or engineer the enterprise and ensure that its intended goals are met, with the result that both researchers and practitioners voiced a need for a comprehensive view of the enterprise in different publications [3; 4; 5; 6; 7; 8].

Many different disciplines contribute to the design of the enterprise, including systems engineering, industrial engineering, information systems, management sciences, psychology, sociology and organisational sciences [1; 9]. The discussions around enterprise design culminates with Giachetti [1, p 3] stating that a *new discipline* is required, called *enterprise engineering* (EE), which could be defined as "the body of knowledge, principles, and practices to design an enterprise". Barjis [10] refines the debate by claiming that EE consists of three subfields namely enterprise ontology, enterprise governance, and enterprise architecture (EA).

EE as a young discipline, which is often regarded as an extension of the fields of industrial engineering or business process management, experiences that "the current status of enterprise engineering initiatives as taken by several universities, is unclear" [11, p 93]. In addition, a plethora of applicable literature and terminology exists from various alternative and associated disciplines, but with a lack of shared meaning [12]. As a consequence there is not a clear description of exactly what EE entails; neither does a research agenda for EE exist, which are both prerequisites for the EE discipline to progress. Before we discuss the research agenda for EE, it is necessary to consider the knowledge that underpins a discipline.

According to Gregor [13], based on Godfrey-Smith [14], the *bodies of knowledge or theories* encompassed in a discipline, need to address *questions related to four classes* namely the: (1) domain; (2) structural or ontological; (3) epistemological; and (4) socio-political.

As a departure point for setting a research agenda for EE, we argue that the *four classes of questions* could also serve as a basis to determine an EE research agenda. The seminal paper on the *discipline of EE*, already provides *EE theories within the classification scheme* (EECS) "as a theoretical research agenda for the enterprise engineering community" [11, p 97]. In this paper, we reason that EECS primarily answers questions pertaining to the *second class*, i.e. the *structural nature* or *ontological character* of theory in EE. We furthermore propose that a research agenda for EE should start with the *first class* of questions, concerning the *domain* of the discipline and suggest that an existing model, the Enterprise Evolution Contextualisation Model (EECM), could be used to define the *domain of the EE discipline*.

The main contribution of this paper is to propose a comprehensive approach towards the establishment of a research agenda for EE, starting with a demonstration of how EECM could be used to answer the *domain* questions pertaining to the EE discipline, namely the *phenomena of interest*; *core problems/topics of interest* and *boundaries* of the EE discipline. Previous research indicated that the existing EE body of knowledge is primarily embedded in multiple enterprise design/alignment/governance approaches [15]. EECM, developed inductively from existing enterprise design/alignment/governance approaches, serves as a common reference model to understand and compare existing EE approaches.

The paper is structured as follows: section 2 provides background theory on *four classes* of questions that are useful in defining the bodies of knowledge or theories of a discipline. We provide background about the constructional components of an existing model, the Enterprise Evolution Contextualisation Model (EECM), since EECM has the ability to answer the *first class* of questions, i.e. the *domain questions* pertaining to the EE discipline. Section 3 applies EECM to define the nature of the EE discipline, i.e. answering three *domain questions* of the EE discipline. In section 4 we discuss future work required in setting a comprehensive agenda for EE research.

## 2 Background Theory

Section 2.1 delineates the *four classes* of questions, as defined by Gregor [13], that are useful in defining a discipline, and we argue that these could guide the definition of the EE discipline as a first step in setting an EE research agenda. In section 2.2, we return to the questions in the *second class*, providing an example of a taxonomy suggested by Gregor [13] to answer questions pertaining to the *second class*. In section 2.3 we present the *EE theories in the classification scheme* (EECS) and reason that EECS primarily answers questions pertaining to the *second class*. Since we propose that the *four classes* of questions are used as an EE research agenda, we suggest that one starts with the first class, namely the *domain* class and present EECM in section 2.4 as a means of defining the domain of EE.

### 2.1 The Bodies of Knowledge Encompassed in a Discipline

Ridley [16, p 12], based on Tardif [17], defines a discipline as a "body of knowledge, definitions, and concepts built up over a long period and receiving consensus recognition by scholars; theories which interrelate the concepts and provide explanations of observed phenomena and permit predictions from them; and well established research methodologies".

In establishing any discipline, Gregor [13, p 611] states that *bodies of knowledge or theories* need to exist to answer question pertaining to *four classes*:

*1. Domain questions.* "What phenomena are of interest in the discipline? What are the core problems or topics of interest? What are the boundaries of the discipline?"

*2. Structural or ontological questions.* "What is theory? How is this term understood in the discipline? Of what is theory composed? What forms do contributions to knowledge take? How is theory expressed? What types of claims or statements can be made? What types of questions are addressed?"

*3. Epistemological questions.* "How is theory constructed? How can scientific knowledge be acquired? How is theory tested? What research methods can be used? What criteria are applied to judge the soundness and rigor of research methods?"

*4. Socio-political questions.* "How is the disciplinary knowledge understood by stakeholders against the backdrop of human affairs? Where and by whom has theory been developed? What are the history and sociology of theory evolution? Are scholars in the discipline in general agreement about current theories or do profound differences of opinion exist? How is knowledge applied? Is the knowledge expected to be relevant and useful in a practical sense? Are there social, ethical, or political issues associated with the use of the disciplinary knowledge?"

### 2.2 Defining the Nature of Theory within a Discipline

Gregor [13] focuses on the information systems (IS) discipline when examining the *second class* of questions, i.e. examining the *nature of theory in IS*. Drawing upon writings from the philosophy of the natural sciences, the social sciences and from the sciences of the artificial, Gregor [13] emphasizes four primary goals of theory for IS:

(1) analysis and description; (2) explanation; (3) prediction; and (4) prescription. Combinations of these goals led to *five types of theory*. Distinguishing features of each *theory type* could be summarised as follows [13, p 620]:

I. Analysis: *Says what is*. The theory does not extend beyond analysis and description. No causal relationships among phenomena are specified and no predictions are made.

II. Explanation: *Says what is, how, why, when, and where*. The theory provides explanations but does not aim to predict with any precision. There are no testable propositions.

III. Prediction: *Says what is and what will be*. The theory provides predictions and has testable propositions but does not have well-developed justificatory causal explanations.

IV. Explanation and prediction (EP): *Says what is, how, why, when, where, and what will be*. Provides predictions and have both testable propositions and causal explanations.

V. Design and action: *Says how to do something*. The theory gives explicit prescriptions (e.g., methods, techniques, principles of form and function) for constructing an artefact.

Similar to the taxonomy of Gregor [13] in defining the nature of IS, Dietz et al. [11] also present a taxonomy or classification scheme to classify EE theories. The next section provides background on the *EE theories in the classification scheme* (EECS), followed by an argument that EECS primarily answers questions pertaining to the *nature of theory in EE*, i.e. focusing on the *second class* of questions that are useful in defining the EE discipline.

## 2.3 Enterprise Engineering Theories in the Classification Scheme (EECS)

Dietz et al. [11] propose a theory-based *methodology* to address enterprise (re-) development in a comprehensive way. In developing the methodology, they propose a theoretical foundation to support three key design concerns: (1) intellectual manageability, (2) organisational concinnity and (3) social devotion. The theoretical foundation is presented in the form of eight theories, categorised according to a classification scheme. Comparable to the *five types of theory* for IS, Dietz et al. [11] present *four classifications of related theories*. Distinguishing attributes and encapsulated theories for each *classification category* could be summarised as follows [11]:

I. Pilosophical: Theories which include philosophical branches of epistemology, phenomenology, logic and mathematics, and assessed in terms of their *truthfulness* within a specific area. The $\varphi$-theory provides a basis for conceptual models in other EE-theories. The $\delta$-theory provides the basis for understanding a system, its processes, events and states. Lastly, the $\tau$-theory provides the basis for understanding the notion of systems, models, and system function vs. construction.

II. Ontological theories: Theories which *analyse phenomena* to identify cause-and-effect and/or predictive relationships, and assessed in terms of their *soundness* and *appropriateness*. The $\psi$-theory defines the ontological essence of organisations in terms of actors that coordinate around the production of services or products. The π-theory explains the construction and operation of technical systems, which involves technical agents (non-human agents).

III. Technological theories: Theories which address means-end relations between phenomena (e.g. designing new ways of addressing existing problems), and assessed in terms of their *rigor* and *relevance*. The β-theory applies *design science* to *design* discrete event systems, which incorporate design-steps such as development, engineering and implementation. The ν-theory is about the construction of a system, using elementary building blocks, such that the removal of an element does not have a combinatorial side effect on the other system elements.

IV. Ideological theories: Theories which address the goals that people may want to achieve in society and specifically within the enterprise, and assessed in terms of their societal *significance*. The σ-theory prescribes a particular governance approach for modern enterprises, co-developing enterprise and employee interests.

Dietz et al. [11] thus present *four classifications of theories* for EE research, whereas Gregor [13] suggests *five theory types* for IS research. Since the EE discipline encapsulates IS [1; 2] it is expected that the *four classifications of theories* for EE, also encapsulate the *five theory types* for IS. Comparing the distinguishing definitions that are provided for the *four classifications of theories* for EE with the definitions of the *five theory types* for IS, interpretive comparisons are possible, as mapped in Table 1:

**Table 1.** Four classifications of theories for EE [11] related to five theory types for IS [13]

| Classifications of theory for EE | Theory types for IS |
|---|---|
| I. Philosophical theories | I. Theory for analysing |
| II. Ontological theories | II. Theory for explaining and III. Theory for predicting |
| III. Technological theories | IV. Theory for design and action |
| IV. Ideological theories | <No mapping> |

Table 1 indicates that the fourth classification category, *ideological theories* for EE, could not be mapped to existing IS theories. *Ideological theories* for EE, addressing the goals that people may want to achieve in society and specifically within the *enterprise*, seems to apply to the EE discipline, but not necessarily to the IS discipline. Further research is required but is not the focus of this paper.

Since the *four classifications of related theories* for EE [11] are comparable to the *five theory types* for IS [13], we argue that the *four classes of related theories* for EE [11] primarily answers the *second class* of questions identified by Gregor [13] for establishing a new discipline.

We propose that a research agenda for EE should start with the *first class* of questions that are required for establishing a new discipline, concerning the *domain* of the discipline and suggest that an existing model, the Enterprise Evolution Contextualisation Model (EECM), could be used to define the *domain of the EE discipline*. The next section presents the background and constructional components of EECM.

## 2.4 The Enterprise Evolution Contextualisation Model (EECM)

Previous research highlighted that fragmentation exists within the EE discipline and the need to provide a common reference model to understand and compare existing knowledge within the EE discipline [18]. Developed inductively from existing enterprise design/alignment/governance approaches, EECM could be used to contextualise/translate an existing approach unambiguously when the approach presented a coherent and consistent *value-creation-paradigm* and a consistent set of *design domains* [15; 18; 19]. Previous research also indicated that a contextualisation model would be useful to extend one approach with elements from another approach when a similar *value-creation-paradigm* exists. De Vries [20] provides an example where the *foundation of execution approach* of Ross et al. [21] was extended by the *essence of operation* approach of Dietz [22]. Other examples of contextualisations (e.g. contextualising the Zachman approach and Open Group approach) are available in previous publications [15; 18].

In this paper, we propose that EECM could also be used define the *domain* of the EE discipline. This section starts with a presentation of EECM, a descriptive model (Fig. 1) that contextualises an existing enterprise design/alignment/governance approach. EECM asks three main questions about a specific approach:

- Question 1: '*Why* should the enterprise use the proposed approach to evolve?'
- Question 2: '*What* should the enterprise evolve?'
- Question 3: '*How* should the enterprise evolve?'

In answering the three questions through a conceptual mechanism, EECM subsequently consists of *four main components* that are presented in the subsequent paragraphs.

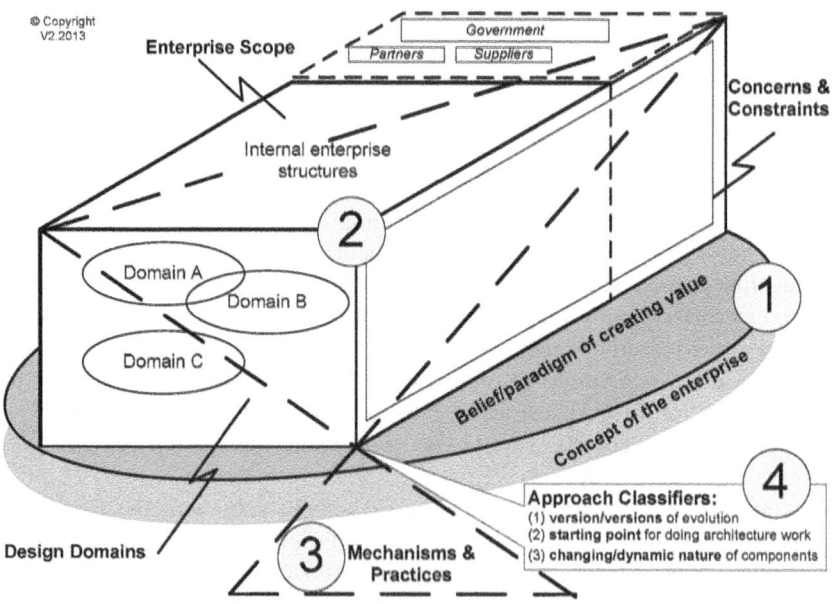

**Fig. 1.** The Enterprise Evolution Contextualisation Model (EECM)

*Component 1: Concept of the Enterprise and Paradigm of Creating Value*
The first component of EECM is presented as the foundation ellipse (Fig. 1, Component 1) [23]. Approach authors used different analogies to define the concept of an enterprise, e.g. machine analogy, biological/uni-minded system analogy or socio-technical/multi-minded system analogy [1; 8; 24]. In addition to the *concept of the enterprise*, approach author(s) also create value propositions in promoting/marketing their approach to possible approach-users. The *paradigm of creating value* thus refers to the philosophical stance of the approach author(s), their way of thinking (see [25]) and their belief-system about what should create value for an enterprise. Together, the *concept of the enterprise* and the *paradigm of creating value*, directs the entire approach for designing/aligning/governing the enterprise. As an example, Dietz et al. [11, p 93-94] indicate that their *concept of the enterprise* is an "organised complexity", a "complex adaptive system" analogous to an "improvisational theatre", whereas their *paradigm of value-creation* is "enterprise (re)-development in an all-encompassing way" [11, p 95].

*Component 2: Dimensions*
The second component of EECM, defines the scope of evolution and consist of three dimensions (Fig. 1, Component 2), represented by three panes of a block: design domains, concerns & constraints, and enterprise scope.

1) Design domains: Literature reveals many different conceptualisations for design domains. Hoogervorst [2, p 134] maintains that the demarcation/delineation of domains reveal "functional or constructional system facets for which design activities are required"; demarcation is not simple and requires specific system knowledge. Dietz [22] delineates design domains as sub-systems for which design activities are required. Yet, defining the boundary of a sub-system is contextual and depends on the intentions of the observer/analyst [1]. As an example, the Open Group [26] defines three design domains: business, information system (which includes application and data), and technology. Another approach author (Hoogervorst [2]) defines four design domains: business (the environmental system, customers requiring products/services), organisation, information, and technology.

Lindström *et al.* [27] highlight four systems that are prominent in enterprise design: the business consumer system, the business organisation system, the ICT system, and the ICT organisation system. Approach authors focus on different levels of scope when designing/aligning/governing the enterprise, some focus on the design of the ICT system within the context of the business organisation system, whereas others focus on the design of the entire enterprise within the context of the business consumer system [15].

2) Concerns and constraints: The concerns are the functional and constructional (non-functional) requirements that need to be addressed during the design of one or more design domains. When the enterprise design domains are demarcated as enterprise sub-systems (i.e. business organisation system, ICT system and ICT organisation system), the concerns include the functional and constructional/non-functional requirements that need to be addressed during the design of the three enterprise sub-systems.

According to Hoogervorst [2] the functional concerns are usually dictated by the user of a system (e.g. the user of the enterprise system), whereas the constructional/non-functional concerns are defined by the system designers. Non-functional concerns not only include required qualities of the designed enterprise (e.g. robustness, agility, flexibility and scalability), but also *constraints*, which are defined by Giachetti [1, p 186] as restrictions imposed "due to physical limitations of resources, due to the environment such as regulatory rules, or due to any reason that justifies defining restrictions on other requirements". Hoogervorst [2, p 298] provides a number of enterprise non-functional *concerns*, e.g. time to market, quality, customer satisfaction, employee satisfaction, employee involvement, safety, costs, compliance and business ethics. Others highlight non-functional concerns, such as governance, security, privacy, workforce, and adherence to standards [9; 28].

3) Enterprise scope: The enterprise scope dimension of EECM reflects the extent of desing/alignment/governance in terms of the internal enterprise structures, such as business units or lines of business, departments, programmes, and projects. Some design/alignment/governance endeavours may extend the boundaries of a single enterprise (i.e. single legal entity) to include design/alignment with external enterprises, e.g. government, partners and suppliers. An example of design across the extended enterprise is the design of a complex supply chain [1]. The structural elements define the boundaries for design/alignment/governance endeavours, and directly influence the required governance responsibilities.

*Component 3: Mechanisms and Practices*
The set of applicable *mechanisms and practices* (Fig. 1, Component 3) that supports a specific design/alignment/governance approach depends on the *concept of the enterprise & belief/paradigm of creating value* (Fig. 1, Component 1) and the design/alignment/governance scope defined by the three *dimensions (design domains, concerns & constraints, enterprise scope)* (Fig. 1, Component 2).

In practice, *mechanisms and practices* are usually organised as an integrated *set of mechanisms and practices* as part of a *methodology*. TOGAF's ADM is an example of a *methodology*, which includes nine sequential and/or iterative phases and numerous mechanism and practices. Hoogervorst [2, p 221, 316] also suggests an alignment *process* to enact alignment on different levels of scope, i.e. enacting design/alignment/ governance across three enterprise sub-systems (business organisation system, the ICT system, and the ICT organisation system). Dietz et al. [11, p 102], in their approach, refer to several *methodologies* in EE.

Nine categories of *mechanisms and practices* (non-exhaustive) were extracted from existing design/alignment/governance approaches [15; 19]. Examples of categories include *architecture description & reference models*; *methodologies*; and *governing principles*.

*Component 4: Approach Classifiers*
EECM provides four classifiers to differentiate between approaches in *how* they ensure consistent design/alignment/governance (Fig. 1, Component 4). The EECM foundation (*concept of the enterprise & belief/paradigm of creating value*) directly influences the design/alignment/governance approach, which in turn influences the set

of *mechanisms and practices* that are required in combination with the design/alignment/governance approach. The approach classifiers are:

*1. Version/versions of evolution:* The version(s) refer to the version of the architecture description with reference to the design domains and concerns. Design/alignment/governance approaches differ in their focus on creating current and/or future versions of architecture description. Some approaches focus on building a complete architecture description (blueprint) of the current (*as-is*) enterprise version. As an example, The Open Group [26] in its ADM follows a systematic process in analysing the current enterprise version to define gaps (gap analyses), prior to defining the future version. The rationale is that the analyses would highlight inefficiencies, reveal opportunities for centralisation or decentralisation, and lead to cost-cutting efforts. Certain approaches focus more on the future (*to-be*) enterprise version, while following a pragmatic approach in building a sub-set of architecture descriptions for the current enterprise version [29; 30; 31].

*2. Starting point for doing architecture work:* With reference to the three primary enterprise sub-systems (i.e. business organisation system, ICT system and ICT organisation system), approaches favour a top-down or bottom-up approach in designing the systems. Some approaches promote a top-down approach (e.g. The Open Group [26]) initially starts with the business organisation system (top level), working towards the ICT and ICT organisation systems (bottom levels)). As an alternative, design could also start at the ICT system (a bottom level). The rationale for starting at the bottom (e.g. using service oriented architecture) is that a flexible IT infrastructure would easily accommodate changes in the business system [32].

*3. Changing/dynamic nature of components:* Enterprise design does not occur at a single point in time, since enterprises evolve over time and are constantly changing [1]. *Dynamics* are at the heart of regulation in organismic systems, rather than control and feedback [33]. Approaches propose different means for addressing the dynamic nature of architecture components. The Open Group [26] maintains that the practice of open standards and boundaryless integration across departmental/divisional/enterprise boundaries address the challenges associated with dynamic changes. Other approaches emphasise governance practices that are required to enact change [34; 35].

Since EECM has been developed inductively (bottom-up) from the existing body of knowledge for designing/aligning/governing the enterprise [15], we argue that EECM represents a categorization and high-level meta-model for the existing body of knowledge within EE. We therefore propose that EECM is used to contribute towards defining the *domain* when defining the nature of the EE discipline, as discussed in the next section.

## 3  Defining the Nature of the EE Discipline

According to Gregor [13], *four classes* of questions are relevant in defining the nature of a discipline. The *four classes* are: (1) domain; (2) structural and ontological; (3) epistemological; and (4) socio-political.

We propose that the four classes of questions are used as a broad agenda for research within the field of EE. When defining the first class, the domain class, Gregor [13] suggests that we consider three questions namely:
1. What *phenomena are of interest* in the discipline?
2. What are the *core problems or topics of interest*?
3. What are the *boundaries* of the discipline?

As mentioned, EECM was developed inductively (bottom-up) from an extensive analysis of current prevalent EE approaches and thus represents a high-level categorisation and meta-model of the existing EE body of knowledge. In the next sections we discuss how EECM could be used to answer the three questions pertaining to the domain of the EE discipline.

### 3.1 The Phenomena of Interest in the Discipline

The era we live in is characterised by rapid changes; the most conspicuous of these are *technological* change, including connectivity; smart devices and ubiquitous computing; and the generation of and access to a vast amount of information. The *enterprise* as a socio-technical system where society meets technology is in the epicentre of the impact of most of these changes. It is plausible to speculate that changes that modern society experience could have an impact on the observed phenomena that characterises the discipline of EE.

During the development of EECM, there was evidence that EE researchers addressed phenomena related to the abovementioned discussion. Phenomena such as the complexity of the enterprise; rapid changing environments impacting on the ability of the enterprise to adapt; the enterprise as a socio-technical *system*; access to information and enterprise knowledge management; the alignment of technology or IT-infrastructure with business strategy; etc. are recurring themes [1; 8; 36; 37; 38].

In our research we observed that there are numerous approaches within EE, but that fragmentation exists with regards to these approaches. Furthermore, there is a distinct lack of a common terminology within EE. In fact, these observations were some of the main motivations for the development of EECM. EECM was constructed through an inductive approach as a meta-model and high-level categorization for all EE approaches at present [15]. The four components of EECM (*concept of the enterprise and belief/paradigm of creating value*; *dimensions*; *mechanisms and practices*; and *approach classifiers* (detailed in section 2.4)) represent high-level categories of phenomena observed by approach protagonists, and the content of the EECM components embody further phenomena of concern. We argue that EECM answers the question of the phenomena of interest, both when we discuss the status quo within EE as motivation for EECM, through the high-level categorisation of EECM components, as well as when the detail content of the EECM components are considered. The phenomena of interest and the core problems are closely related and are thus further discussed in the next section.

## 3.2 The Core Problems or Topics of Interest

When observed phenomena within any discipline are analysed, distinct domain problems are identified, leading to the second question regarding the domain of the discipline: *what are the core problems or topics of interest within the EE discipline*?

In the development of EECM, it was clear that the plethora of approaches and definitions, as well as conflicting terminology, is one of the problems in the EE domain, which EECM attempts to resolve. Furthermore, the three main questions of EECM that guide the contextualisation of an approach, namely: '*Why* should the enterprise use the proposed approach to evolve?'; '*What* should the enterprise evolve?'; and '*How* should the enterprise evolve?' could be considered as meta topics of interest within the EE discipline.

When analysing and contextualising an approach using EECM, several more detail topics of interest for the EE discipline are identified. For example, a number of *value-creation paradigms* could be extracted from existing EE approaches, each focusing on a different problem or topic of interest. The following topics (non-exhaustive list) emerged from EE approach literature consulted:

1. Enterprises do not have an *aggregate view* of the enterprise to direct its evolution. A *value-creation paradigm* is thus to provide an *aggregate view* for directing the enterprise in terms of required high-level processes and IT capabilities [21; 39; 40]. Others [31] also emphasise the intention of directing the enterprise on a strategic level; i.e. creating a *holistic view* of the business processes, systems, information, and technology, which would lead to more intelligent investment decisions.

2. There is a lack of describing the enterprise components, their interaction and interrelationships in a consistent way to ensure holistic solutions in terms of the solution components [26; 30; 40; 41; 42; 43]. A *value-creation paradigm* is thus to create a *systems view*, i.e. the "fundamental organisation of a system, embodied in its components, their relationships to each other and the environment, and the principles governing its design and evolution" [44]. A *systems view* should focus on reducing complexity of IT and business processes across the breadth of the enterprise, making a company more agile [31].

3. All enterprises face the need to continuously transform from an existing state to a future state. A *value-creation paradigm* is to enable transformation from a current state to a future state, i.e. translating business vision and strategy into effective enterprise change [9; 28; 30; 45; 46].

4. Enterprises still fail to implement strategic initiatives successfully, which is primarily due to the lack of coherence and consistency among the various components of the enterprise [2; 11]. A prominent *value-creation paradigm* is thus *enterprise governance*, i.e. key principles that are required to govern the design and evolution of the enterprise. Although many approach authors focus on IT governance [26; 30; 34; 40; 42; 46], others [2; 11] take and enterprise-wide governance view.

5. Large enterprises experience that multiple decision-makers are involved during enterprise design, each with their own interests. A *value-creation paradigm* is thus to provide an integrated and transparent *representation of all interests* and their current state of alignment [47].

### 3.3 The Boundaries of the Discipline

As indicated by previous research, EECM is successful at present when used to contextualise any prevalent EE approach [15]. Being a non-prescriptive model, EECM highlights the commonalities within the existing body of EE knowledge, but also acknowledge different stances towards the design/alignment/governance of the enterprise. EECM as a meta-model thus represents the boundaries of the current EE discipline as discussed in the following sections.

*Component 1: Concept of the Enterprise and Paradigm of Creating Value*
Since the enterprise is an artificial man-made entity [1], different conceptualisations exist for defining the enterprise and its components. The *concepts* and analogies that are used in *defining an enterprise*, together with the *belief/paradigm of what could create value* in an enterprise constrain and direct the entire EE approach. The *concept of the enterprise & paradigm of creating value* thus provides a philosophical boundary for the EE discipline, conceptualising about people and artefacts that work towards common enterprise goals. The philosophical boundary thus excludes concepts of artefacts/systems that do not contribute towards the realisation of common enterprise goals. Conceptualising about the structure of plants would thus be excluded from the domain of EE, unless used to organise people and artefacts/systems around common enterprise goals, e.g. using the characteristics and dimensions of plants/crops to organise the production and delivery of crops to customers.

*Component 2: Dimensions*
The boundaries of EE are further defined by the three dimensions of EECM. Hoogervorst [2, p 134] maintains that the demarcation/delineation of the first dimension (*design domains*) reveal "functional or constructional system facets for which design activities are required". Many different demarcations of enterprise design domains exist in literature [18]. Domain experts apply their domain-specific knowledge in addressing the second dimension (functional and non-functional *concerns* and existing enterprise *constraints*) during the design process. Finally, the third dimension (*enterprise scope*) acknowledges existing organising structures within the enterprise system, which divides an enterprise into manageable parts, but also creates integration challenges and consistent evolution of the various organisational parts. The three *dimensions* of EECM thus define a design-scope boundary for the EE discipline.

*Component 3: Mechanisms and Practices*
Several categories of *mechanisms and practices* have emerged from literature, which demarcate the EE discipline in terms of the relevant and supporting mechanisms and practices necessary for the chosen design domains. The *mechanisms and practices* of EECM ensure that practical facets are included for the design-scope boundary of the EE discipline.

*Component 4: Approach Classifiers*
Three *approach classifiers* demarcate the EE discipline further to distinguish different patterns within existing EE approaches. Inductive research highlighted that an EE approach usually has a preference for (1) a *specific version of evolution* (current or

future version of enterprise architecture description), (2) *starting point for doing architecture work* (top-down or bottom-up), and (3) a strategy to *address the dynamic nature of enterprise components.*

## 4 Conclusion and Future Research

To conclude, we support the arguments that there is a need for an EE discipline. We furthermore support the notion that a research agenda for EE should be developed to ensure that research is directed and that the body of knowledge evolve in a systematic and scientific way. When defining the research agenda for a discipline it is necessary to understand the bodies of knowledge or theories that underpin the discipline. Gregor [13, p 611] states that bodies of knowledge or theories need to exist to answer question pertaining to *four classes* namely the domain, structural or ontological, epistemological and socio-political. We argue that this approach could be used to assist with defining the research agenda of the EE discipline. Yet, we also acknowledge that future research would be required to validate the *four classes of questions,* adding more rigour to Gregor's suggested classes of questions.

Since the seminal paper on the *discipline of EE,* already provides *EE theories within the classification scheme* (EECS) "as a theoretical research agenda for the enterprise engineering community" [11, p 97], this paper compared EECS with the *four classes of questions* posed by Gregor for establishing any discipline. Using Gregor's *four classes of questions* as a departure point, we argued that the EECS primarily answers questions related to Gregor's [13] *second class of questions*, namely the *structural and ontological questions* of the EE discipline. Since our study was only *limited* to an analysis of the EECS, we propose for further research an extended study, which incorporates all referenced literature in the seminal paper on the *discipline of EE* of Dietz et al. [11] to assess whether the referenced literature answers the *four classes of questions* presented by Gregor [13].

We conclude that EECM could be adopted as a mechanism for answering the *first class of questions* regarding the *domain* of the EE discipline, namely *what phenomena are of interest in the discipline?*; *what are the core problems or topics of interest?*; and *what are the boundaries of the discipline?* Supplemental research is also suggested to further demarcate the *domain* of the EE discipline.

Finally, further research should address setting the complete research agenda, thus also considering the *epistemological* and *socio-political questions.*

## References

1. Giachetti, R.E.: Design of Enterprise Systems. CRC Press, Boca Raton (2010)
2. Hoogervorst, J.A.P.: Enterprise Governance and Enterprise Engineering. Springer (2009)
3. Towill, D.R.: Successful Business Systems Engineering Part I: The Systems Approach to Business Processes. Engineering Management Journal 7(1), 55–64 (1997)
4. Rouse, W.B.: Embracing the Enterprise. Industrial Engineer 36(3), 31–35 (2004)
5. Liles, D.H., Johnson, M.E., Meade, L.: The Enterprise Engineering Discipline. In: Proc. of the Society for Enterprise Engineering, Orlando, FL (1995)

6. Martin, J.: The Great Transition: Using the Seven Disciplines of Enterprise Engineering to Align People, Technology, and Strategy. American Management Association, New York (1995)
7. Kappelman, L.A., McGinnis, T., Pettit, A., Salmans, B., Sidorova, A.: Enterprise Architecture: Charting the Territory for Academic Research. In: Kappelman, L.A. (ed.) The SIM Guide to Enterprise Architecture, pp. 96–110. CRC Press, Boca Raton (2010)
8. Van Tonder, C.L., Roodt, G.: Organisation Development: Theory and Practice. Van Schaik Publishers, Pretoria (2008)
9. Bernard, S.A.: An Introduction to Enterprise Architecture EA3, 2nd edn. Authorhouse, Bloomington (2005)
10. Barjis, J.: Enterprise Modeling and Simulation within Enterprise Engineering. Journal of Enterprise Transformation 1(3), 185–207 (2011)
11. Dietz, J.L.G., Hoogervorst, J.A.P., Albani, A., Aveiro, D., Babkin, E., Barjis, J., Caetano, A., Huyments, P., Iijima, J., Van Kervel, S.J.H., Mulder, H., Op't Land, M., Proper, H.A., Sanz, J., Terlouw, L., Tribolet, J., Verelst, J., Winter, R.: The Discipline of Enterprise Engineering. International Journal of Organisation Design and Engineering 3(1), 86–114 (2013)
12. Lapalme, J.: 3 Schools of Enterprise Architecture. IT Professional 13(6), 1–7 (2011)
13. Gregor, S.: The Nature of Theory in Information Systems. MIS Quarterly 30(3), 611–642 (2006)
14. Godfrey-Smith, P.: Theory and Reality. University of Chicago Press, Chicago (2003)
15. De Vries, M., Van der Merwe, A., Gerber, A.: Towards an Enterprise Evolution Contextualisation Model. In: First IEEE SMC Conference on Enterprise Systems (ES 2013), IEEE Explore (2013)
16. Ridley, G.: Characterising Academic Information Systems in Australia: Developing and Evaluating a Theoretical Framework. Australian National University E Press, Australia (2008)
17. Tardif, R.: The Penguin Macquarie Dictionary of Australian Education. Penguin, Victoria (1989)
18. De Vries, M.: A Process Reuse Identification Framework Using an Alignment Model. PhD Thesis. University of Pretoria, Pretoria (2012)
19. De Vries, M.: A Framework for Understanding and Comparing Enterprise Architecture Models. Management Dynamics 19(2), 17–29 (2010)
20. De Vries, M.: Using a Classification Schema to Compare Business-IT Alignment Approaches. International Journal of Industrial Engineering: Theory, Applications and Practice 20(3-4), 298–308 (2013)
21. Ross, J.W., Weill, P., Robertson, D.C.: Enterprise Architecture as Strategy: Creating a Foundation for Business Execution. Harvard Business School Press, Boston (2006)
22. Dietz, J.L.G.: Enterprise Ontology. Springer, Berlin (2006)
23. De Vries, M., Gerber, A., Van der Merwe, A.: Engineer the Engerprise, https://sites.google.com/site/engineertheenterprise/
24. Gharajedaghi, J.: Systems Thinking: Managing Chaos and Complexity, 3rd edn. Elsevier, Burlington (2011)
25. Wijers, G.M., Heijes, H.: Automared Support of the Modelling Process: A View Basedon Experiments with Expert Information Engineers. In: Steinholtz, B., Sølvberg, A., Bergman, L. (eds.) CAiSE 1990. LNCS, vol. 436, Springer, Heidelberg (1990)
26. The Open Group: TOGAF Version 9.0, Enterprise Edition (2009), https://www.opengroup.org/online-pubs

27. Lindström, A., Jonhnson, P., Johansson, E., Ekstedt, M., Simonsson, M.: A Survey on CIO Concerns - Do Enterprise Architecture Frameworks Support Them? Information Systems Frontiers 8(2), 81–90 (2006)
28. Schekkerman, J.: How to Survive in the Jungle of Enterprise Architecture Frameworks, 2nd edn. Trafford Publishing, Victoria (2004)
29. Buchanan, R.D., Soley, R.M.: Aligning Enterprise Architecture and IT Investments with Corporate Goals. OMG and Meta Group, Needham (2002)
30. Lapkin, A.: Gartner Clarifies the Definition of the Term 'Enterprise Architecture'. Report ID: G00156559, Gartner Research, USA (2008)
31. DeBoever, L.R., Paras, G.S., Westbrock, T.: A Pragmatic Approach to a Highly Effective Enterprise Architecture Program. In: Kappelman, L.A. (ed.) The SIM Guide to Enterprise Architecture, pp. 156–161. CRC Press, Boca Raton (2010)
32. Robertson, B.: Architecting the Technology Viewpoint Requires Defining a Service-Oriented Infrastructure Strategy. Report ID: G00134007, Gartner Research, USA (2005)
33. Hitchins, D.K.: Advanced Systems Thinking, Engineering and Management. Artech House, Boston (2003)
34. Wagter, R., van den Berg, M., Luijpers, J., van Steenbergen, M.: Dynamic Enterprise Architecture: How to Make it Work. John Wiley & Sons Inc., New Jersey (2005)
35. Bittler, R.S., Kreizman, G.: Gartner Enterprise Architecture Process: Evolution 2005. Report ID: G00130849. Gartner Group, USA (2005)
36. Rebovich, G., White, B.E.: Enterprise Systems Engineering. CRC Press, Boca Raton (2011)
37. Zachman, J.A.: Z101 Master Class: Framework Foundations. Report, Zachman International (2009)
38. Luftman, J., Ben-Zvi, T.: Key Issues for IT Executives 2010: Judicious IT Investments Continue Post-Recession (2012), http://blog.cionet.com/wp-content/uploads/2010/12/MISQE-2010-Key-Issues-for-IT-Executives-2.pdf
39. Boar, B.H.: Constructing Blueprints for Enterprise IT Architecture. J. Wiley, New York (1999)
40. Winter, R., Fischer, R.: Essential Layers, Artefacts, and Dependencies of Enterprise Architecture. Journal of Enterprise Architecture 3(2), 7–18 (2007)
41. EA Research Forum: Enterprise Architecture Definition. Report, Enterprise Architecture Research Forum, Pretoria (2009)
42. Theuerkorn, F.: Lightweight Enterprise Architectures. Auerbach Publications, New York (2005)
43. Handler, R.: Enterprise Architecture is Dead - Long Live Enterprise Architecture (2007), http://itmanagement.earthweb.com/columns/article.php/11079_3347711_1
44. ISO/IEC JTC 1/SC 7 committee: ISO/IEC/IEEE 42010 Systems and Software Engineering - Architecture Description (2012), http://www.iso-architecture.org/42010/docs/ISO-IEC-FDIS-42010.pdf
45. GAO: Enterprise Architecture: Leadership Remains Key to Establishing and Leveraging Architectures for Organisational Transformation (2008), http://www.gao.gov/new.items/d06831.pdf
46. Willis, T.: Enterprise Architecture Workshop Notes. Report, Gartner Research, Pretoria (2009)
47. Sidorova, A., Kappelman, L.A.: Enterprise Architecture as Politics: An Actor-Network Theory Perspective. In: Kappelman, L.A. (ed.) The SIM Guide to Enterprise Architecture, pp. 70–88. CRC Press, Boca Raton (2010)

# What Does DEMO Do? A Qualitative Analysis about DEMO in Practice: Founders, Modellers and Beneficiaries

Céline Décosse[1], Wolfgang A. Molnar[1], and Henderik A. Proper[1,2]

[1] Public Research Center Henri Tudor, 29 Avenue Kennedy, L-1855 Luxembourg, Luxembourg
[2] Radboud University, Comeniuslaan 4, 6525 HP Nijmegen, The Netherlands
{celine.decosse,wolfgang.molnar}@tudor.lu, e.proper@acm.org

**Abstract.** Our goal in this exploratory study is to gain insights about the actual use of DEMO. As we aim at understanding how the use of DEMO influences its context of use and is influenced by it, the study is based on a qualitative approach. 13 stakeholders acquainted with DEMO were interviewed. As DEMO is an artefact, design science literature is relevant to reflect upon the observation of DEMO in practice. We investigated and analysed the views of DEMO founders, DEMO modellers and DEMO beneficiaries about DEMO definition, purpose and scope, results, ease and context of use. We used a subset of criteria of progress for information systems design theories to observe DEMO. Interview results are then exposed and analysed.

**Keywords:** DEMO in practice, qualitative interviews, design science evaluation criteria, artefact observation.

## 1 Introduction: Motivations and Research Questions

In this paper we are concerned about the investigation of DEMO (Design and Engineering Methodology for Organizations [1]) in practice, from various types of stakeholder's points of view. "In practice" refers to the actual use of DEMO in defined contexts, as opposed to the intended use of DEMO modeller or the expected use of DEMO founders, which are out of the scope of this paper. Sometimes called "a methodology" [1–3] or a "method" [2] in the literature, "a formal language and definitely a way of thinking" by its users, "a way of thinking, a way of understanding" by a DEMO founder, DEMO offers a set of axioms, thinking patterns and graphical models that allow its users to produce concise models of organizational processes. The application of DEMO seemed to be very promising in some projects. For example, DEMO is said to have helped to "construct and analyse more models in a shorter period of time" [3] p10. Therefore, we were curious about the performance of DEMO in practice. In addition, we had access to DEMO practitioners who would agree to have the projects where DEMO was applied investigated by researchers.

Our motivation in this paper is exploratory: we investigate the actual use of DEMO from stakeholders' perspective to know how DEMO is seen from field people. The study is based on a qualitative approach: we collected data with 13 semi-structured interviews and analysed them using an interpretive approach. This paper reports about this analysis. The questions addressed in this paper are the following. From stakeholders' point of view: What is DEMO? What is DEMO useful for and not useful for? Who are DEMO beneficiaries? What are the results of using DEMO? Is DEMO easy to use? What parts of DEMO are useful?

The original contributions of this paper are the insights we gained about DEMO in practice from stakeholders who have been acquainted with DEMO for years. The paper is structured as follows: in the introduction, we motivated our study and defined the questions to be addressed. Section 2 is a short literature review about already existing investigations concerning the actual use of DEMO. In section 3, we present the research design. It includes the research approach, the theoretical basis for data collection and analysis and facts about the actual data collection. Section 4 presents the actual data analysis. Section 5 presents some research contributions and bias and the conclusion.

## 2  Previous Investigations Concerning the Use of DEMO

In [4], we performed a literature review investigating whether DEMO had been evaluated in practice. We found two papers dealing with a partial evaluation of DEMO in practice across several cases: "The first one [27] focuses on the adoption of DEMO by DEMO professionals in practice in order to improve this adoption. The second one [11] investigates DEMO as a means of reflecting upon the Language/Action Perspective; the DEMO related part of this paper aims at finding out how the actual application of DEMO differs from its intended application". Both studies only took into account DEMO professionals views. So, as far as we know, no study has been performed yet with the goal of exploring the use of DEMO in practice by a variety of stakeholder types. In [4], the research effort was focussed on the interpretive research approach and on the relevance of investigating design sciences to observe a method. Definitions of observation criteria and details about papers [27] and [11] were given. Besides, preliminary promising results of the interviews were given whereas the complete analysis of the interviews had not been performed at that time. Alternatively, the current paper is focused on interviews analysis: the coding process is partly exposed and interviews analysis results are reported upon.

## 3  Research Design

### 3.1  An Interpretive Approach for Exploring the Use of an Artefact

As we are aware that DEMO stakeholders have their own assumptions, beliefs and perceptions and that they construct realities through social interaction, we used a qualitative research approach to "capture data on the perceptions of local participants"[5] p7. To produce "an understanding of the context of the information system, and the process whereby the information system influences and is influenced by the context" [6] p4-5, we followed Walsham in adopting an interpretive approach for exploring the use of DEMO [4] p6.

This approach consists in interviewing stakeholders with an interview guideline structure based on the criteria of progress for Information Systems (IS) design theories proposed by Aier and Fischer [4, 7]. This approach is purposely not DEMO specific. Indeed, whether DEMO being viewed as a method, a methodology, a way of thinking or a modelling language, we consider DEMO as being an artefact that is designed, performed and evaluated by human people. DEMO was created and originally applied in the context of information system engineering or reengineering. Later, it has been used for organizational analysis purposes. For these reasons, we found it relevant to explore the design science literature to define the interview guideline's themes. In order to produce results that are "credible, dependable and replicable in qualitative terms" [5] p5, we expose the way we worked to perform the study in the following paragraphs.

### 3.2 Theoretical Basis for Data Collection

**A Priori Conceptual Framework, Theme, Items and Questions Elaboration**
In this study, the only instrumentation employed to collect data are interviews about DEMO. We elaborated a conceptual framework in which we gathered themes we wanted to study. They come from a literature review about design science artefacts and method evaluation, from some stakeholders feed-back about DEMO and from a brainstorming with fellow researchers. This a priori conceptual framework (Fig. 1) is our "map of the territory to be investigated" [5], p20.

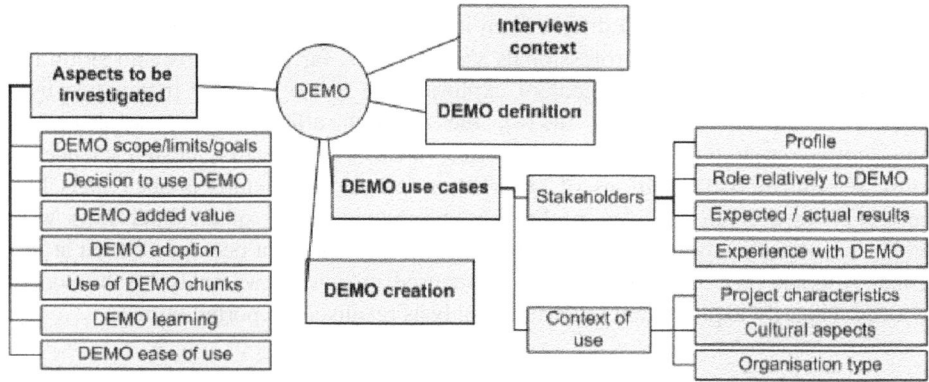

**Fig. 1.** A priori conceptual framework: research themes to be investigated

For each theme, we created a set of items – about 95 in total: 85% concerning the practical use of DEMO and 15% aiming at investigated the creation of DEMO by its founders (context, goals). The latter ones are almost not reported about in this paper because it is focussed on the actual use of DEMO. For each item, we formulated a set of questions both to cover various aspects of the item and various ways of considering the question from stakeholders' perspective. As for themes definition, these questions were elaborated with fellow researchers and come from their experience and a design science literature review. In particular in [4] we reflected upon Aier and Fischer's criteria of progress for information systems theories [7]. To reflect upon criteria of progress for design theories, they investigated artefact evaluation criteria in design science literature. We applied the results of their reflection the other way round: we used some of their criteria of progress for information systems theories as themes of the interview guideline[1] to observe an artefact: DEMO. Due to space limitation, we do not explain here the details of the reasoning. Resulting interview themes are the following: usefulness and efficiency, simplicity – which includes the ease of use, broadness of purpose and scope and external consistency. We consider the "broad purpose and scope" as being "the capacity of adapting the artefact to different purposes and scopes without its usefulness decreasing" [7] p154. Following [7], we considered that the external consistency of an artefact corresponds to the notion of "fidelity with real world phenomena" proposed by March and Smith's [8].

In order to capture the experience of the individuals with DEMO during the interviews, interviewers tried to avoid "leading the witness" by asking open questions of several types (how, what, when, who, what for, etc.)

**Actual Data Collection through Semi-structured Interviews**

*Interviewees sampling.* We wanted to interview multiple stakeholder types: DEMO founders, DEMO professionals and modellers (who produce DEMO models) and DEMO beneficiaries (people involved in projects where DEMO was applied, e.g. project managers). We then used a purposive snowball sampling technique [5] p30. As a result, we interviewed 13 persons. They were free to accept or refuse the interview; there was no obligation from their organizations. One interviewee only had to ask for the authorization to be interviewed from his organization. All interviewees enthusiastically agreed to have their projects investigated by researchers.

*Interviews setup.* Interviews were conducted individually and took place either at the interviewee workplace or in a conference room. One was performed by Skype. All interviewees agreed to be audio-recorded and almost all to be cited. During interviews, we freely added, adapted or removed questions according to the already provided information and to interviewee's knowledge [9]. Interviewers endeavoured to provide a friendly listening without giving their own opinions so as to encourage interviewee to speak freely and to have less bias in the answers [10]. One of the interviewers, Niek, is a Dutch DEMO expert. He is at the board of the enterprise institute which promotes DEMO. The second interviewer, Céline, is a French business analyst whose knowledge about DEMO could be summarised in a few short lines. Céline usually played the part of the main interviewer and Niek acted as a "shadow" interviewer: he observed, asked additional questions and took notes. For practical reasons, interviews were performed in English.

---

[1] Available at www.ee-team.eu/repository/
celine-decosse/Guideline_What-does-DEMO-do.pdf

*Collected data.* Right after each interview, interviewers wrote down "interview highlights" notes about what had been said or the interviewee's reactions to specific subjects. Average time of interviews is one hour an thirty minutes. Twenty hours of recordings have been collected and transcribed by the interviewers. Interviews took place from May to July, 2012.

## 3.3 Theoretical Basis for Data Analysis

This paragraph exposes the method we applied to get from pages of transcriptions to the final conclusions. Interview transcripts were coded two times by the same researcher, after all interviews have been conducted. Due to the fact that coding is a selection process, coding is considered as a part of the analysis [5] p72. We used qualitative data analysis methods proposed by Miles, Huberman and Saldaña [5] p7.

**An inductive Mode for the First Cycle Coding**
First cycle coding goal is to summarize segments of data [5]. During this process, we followed an inductive approach: although we kept in mind the conceptual framework [5] p21 that we had defined keep our research effort focused, we endeavoured to pay attention to recurring elements, elements related to roles and interactions between people, motivations or social rules [5] p19 or pieces of text that may help us to have "an overview of the context under study: its social arrangement, its ways of working, its explicit and implicit rules" [5] p9 and [11]. Indeed, observing DEMO in practice includes observing the context of use of DEMO. The inductive mode allows the emergence of unexpected themes. About 150 codes emerged.

**Second Cycle Coding: Pattern Codes**
Pattern coding is a way of grouping first cycle codes into a smaller number of categories, themes or constructs. Pattern codes usually consist of four, often interrelated, summarizers: categories or themes, causes/explanations, relationship among people, theoretical constructs [5] p86-87. To define the pattern codes, we mainly grouped the already existing codes by similarities, moved and merged sub-codes and created new codes to gather sub-codes in.

## 3.4 Interviewees' Profiles

13 people have been interviewed, from whom 12 are Dutch speaking males and one is a Portuguese male. 12 interviewees were aged between 42 and 62 when interviewed. A DEMO founder (Jan Dietz) was 67. Interviewee's level of English was either good enough to allow them to answer question with nuances and details or, most often, fluent. All research participants have a technical or scientific background or current position: civil engineering (9), mathematics, information technology or systems. They are independent consultants (5), DEMO professionals (9) or acquainted with DEMO (3), researchers (6) or managers (6), professor in information systems (4) – these sets are overlapping. 4 interviewees are members of the enterprise engineering institute board. Several interviewees (3) have founded their own company. Interviewees' level

of education varies between 4 to at least 8 years for researchers. To sum up, interviewees are all highly educated, have high competence profiles, are used to making decisions, are acquainted with modelling and are interested in having strategic views. Some of them participated in the foundation and definition of DEMO. We cited them as "a founder" whereas other interviewees are cited as X1 to X11.

## 4 Data Analysis

This sections reports about the interviews analysis. We do not mention stakeholder's types unless theirs views diverge. Indeed, for the themes we chose to explore in this paper, stakeholders views tend to converge and if they diverge, the stakeholder type is usually not a discriminant. Besides, we indicate when stakeholders views converge, are not contradictory, or diverge. First or all, interviewees and projects are presented.

### 4.1 Projects in Which DEMO Was Applied

Whereas interviewees mentioned about 20 situations in which DEMO have been applied, most of them focused on only two projects during the interviews: VISI and Air France KML Cargo Information Technology (IT) merge. VISI stands for "creating conditions (V) for the introduction (I) of standardization (S) of ICT (I)." VISI concerns the ground, road and water building sector or the sector that performs infrastructure projects for the transport by road or water. VISI development project with DEMO took place from 1998 to 2004 and resulted in the VISI standard, an IT supported framework aiming at improving communication between construction project partners by regulating the exchanges between parties. In Air France KML Cargo IT merge project, DEMO intervention lasted six weeks, from April to June 2008 [2]. DEMO was used to analyse Air France Cargo ICT and KLM cargo ICT in order to allow a co-operation between those ICT systems.

### 4.2 What are Interviewees Talking about When They Say "DEMO"?

It seems that DEMO professionals, including a founder, consider DEMO as a way of thinking and as a way of modelling an organization with a modelling language whose constructs are based on the psi-theory. Then, when answering "a way of thinking" to the question "What is DEMO?", they may actually refer to the "psi-theory" [1]. Some non-DEMO professionals who worked on the VISI project say that they tend not to distinguish precisely between the added-value of DEMO and the added-value of the VISI project results. As some of them have not been in contact with DEMO for 8 years, their contribution to our study is more precious regarding the overall experience they had with DEMO than regarding DEMO or DEMO scope definitions.

## 4.3 DEMO Definition – Convergence and Divergence

**DEMO is a Way of Thinking That Comes with a Way of Modelling – Convergence**
Interviewees' opinions converge about DEMO being "a way of thinking that comes with a way of modelling. The following quotes were issued by stakeholders of various stakeholder types: DEMO is "a formal language and definitely a way of thinking, yes!", "It is a very good tool, a way of thinking to produce the VISI standard", "It is actually a way of thinking. Not more", "I've never known the distinction between the methods and the methodologies but I think it's a way of thinking". Besides, interviewees think that applying the way of thinking without using the modelling is still using DEMO. Interviewees did not express precisely the relationship and limits between DEMO and the psi-theory – they were not asked to; however it seems, for interviews who spontaneously mentioned the psi-theory, that they see it as DEMO underlying theory and part or all of its way of thinking.

**Is DEMO a Method? – Divergence**
*DEMO was not designed to be a method.* A DEMO founder explains: "it was not my idea in fact to develop a methodology. [...] But the main thing I developed (...) is the psi-theory, a theory about the operation and construction of organizations. "Methodology" is just a word to indicate that (...) DEMO has an underlying theory, whereas a method normally does not need to be founded on a theory."

*With the "5 ways" method definition: DEMO is seen as a method – or not.* To explain why they see DEMO as a method or not, several interviewees refer to the view of Seligman et al. [12] on information systems methodologies, for whom methodologies are characterised with "5 ways": the "way of thinking", the "way of working" (how to do things), the "way of controlling" (how to manage things), the "way of modelling" and the "way of supporting" (tools). Interviewees who refer to Seligmann et al. method definition either do it to explain that DEMO is a methodology because it has a way of thinking (X8), or on the contrary argue that DEMO cannot be called a method because of the lack of way of working, "the poor tooling" and some weaknesses in the way of modelling. A DEMO founder said: "for the way of working, the way of project management and also supporting.... well, there is much of improvement to be done".

*Without the "5 ways" method definition: DEMO is a method – or not.* Interviewees who do not refer to Seligmann et al. method definition say: "DEMO is a method with a built-in language inside but it could have been used with another language, it's mainly a method", "I see DEMO as a powerful method of modelling transactions and information, not as a language" or explain that DEMO is "a descriptive methodology" whose use has edge effects: "people working with DEMO start talking a language about the ontological, infological levels or actor roles".

## 4.4 DEMO Purpose and Scope – Convergences and Divergences

**Various Subjects Where Interviewee's Views Converge**

*DEMO, a power-free business modelling tool.* Interviewees converge in thinking that DEMO has a huge value for organizational analysis and modelling. X10 and X3 add: "DEMO has nothing to do with power; it has to do with analytical thinking." They meant that DEMO models and way of thinking do not reflect – respectively take into account – an organization's power distribution, whether being a formal one (organization chart) on an actual one (who decides in practice).

*DEMO is suited when people need a high-level view of an organization.* X4 says that DEMO can be used "in situations where people have lost the oversight." For X3, DEMO "shows the big picture" and DEMO models are appropriate for designing domain models. As such, DEMO can be used as an instrument for business/IT alignment. For a DEMO founder, DEMO bridges information systems and organizational sciences. Other interviewees support this view when explaining how successful DEMO is to analyse a business in order to design or re-design its IT or non-IT implementations.

*DEMO is valuable in case of organizational change and enterprise transformation.* A DEMO founder says: "DEMO is applicable for any enterprise change. Transformation is used for big changes. It is for any." X3 explains how they used DEMO for this: "We built scenarios and we mapped them back to the DEMO model: "This transaction is performed by three different departments with three different tools but this is actually the same transaction", which is from the change perspective or from the organizational perspective a very very valuable information." X5 experienced that, as an organizational analysis tool, DEMO can help in providing informed governance for complex enterprise transformations.

*DEMO does not help in scoping the problem area.* Interviewees are unanimous: DEMO enables its users to have a specific view on an area of concern, however, the scope of this area – the problem area – has to be determined beforehand, even if, as X9 stated, "DEMO allows you to decompose the process and the decomposition of course is important to find yourself a new concentration target focus, part of the problem that you need to solve".

*Various limits of DEMO: what is DEMO not useful for (but not meant at).* For interviewees, DEMO suffers from a lack of bridges towards various types of implementation: X6 says that although DEMO is a good analysis tool, it does not help in developing IT systems in terms of code generation. X1 explains that whereas DEMO applies to systems, it does not take into account the physical situation (e.g. localisation) of the system. X3 underlines that "DEMO completely lacks political thinking. It's not suited for that, which is a large part of what happens in an organization. I think it's both a lack and strength actually." X5 suggests to create a DEMO add-on bearing prescriptive recommendations to evaluate alternative organizational implementation solutions for a DEMO model for cases such as e.g. "Do we put these actor roles in a shared service centre, outsource them or let them still be part of the same organizations?". He also advises to combine DEMO with the notion of quality of service.

**Is DEMO Prescriptive or Descriptive? – Convergence and Divergence**
Interviewees' answers to this question depend both on whether they consider DEMO as a modelling language or as a tool for analyzing organizations and on the meaning they assign to the terms "prescriptive" and "descriptive".

*As a way of modelling, DEMO is seen as prescriptive... or not – Divergence.* An interviewee explains: "No it is not prescriptive, it is descriptive but there is a very strict recipe for the description, and that is a key difference." Some other interviewees see DEMO as being prescriptive because of (a) the novelty of DEMO's way of thinking for them: DEMO "forces" its new users to look at their scope of interest with constructs that are unusual for them (an interviewee mentioned "a prescriptive way of thinking"). The idea of seeing DEMO as being prescriptive because of its novelty is supported by the fact that the more interviewees are used to working with DEMO, the less they tend to see DEMO as being prescriptive; (b) the formality of its constructs: DEMO modelling constructs are coercive and then perceived as prescriptive and (c) of the distinction axiom: once the scope of interest has been defined by the modeller, DEMO tries to enforce, through the distinction axiom, the selection of elements to be modelled within this scope [1].

*DEMO, a normative (but not prescriptive) tool for thinking up organizations DEMO - Convergence.* A DEMO founder explains "DEMO doesn't solve things. It is basically a way of thinking, expressed in models. […] It does not tell you what you should do, it helps you in making decisions" because the models provide a better understanding of the world: "by looking at the world in some way, by some theory, I do not change the world, I see it differently. You could say it is somehow normative because you now understand organizations as networks of actors and transactions. (…) So it is not prescriptive in the way that I tell the organization: now you have to do it in this way". Two interviewees pointed out that although DEMO is actually descriptive, it has a high potential as a prescriptive approach for organizational and IT implementations.

### 4.5 DEMO Beneficiaries Are Architects, Not Implementers – Convergence

Interviewees mentioned enterprise architects, domain architects, business architects as DEMO beneficiaries; X8 recalls that the construction diagram has been taught to quality assurance people. Interviewees converge in thinking that IT implementation oriented people as projects architects may not understand DEMO. Several interviewees suggest that having an engineering background may help; one of them explains: "People with a financial background (…) don't see a design problem when it hits them. So [DEMO modellers] are all mathematicians, of the engineering type, biologists by background." Another one says: "DEMO is really suited for business architects, not management. I am an exception."

## 4.6 DEMO Use Outcomes, Added-Value and Conditions to Achieve Them – Convergence

Our research participants consider DEMO as a tool to reflect upon the communication between responsible parts of a business process. By offering a set of models, DEMO allows its users to model their business processes in terms of transactions and responsibilities. According to interviewees, using DEMO seem to (a) quickly provide a mental or graphical picture of an organization's business processes (composed by one or several DEMO models) which displays transactions and responsibilities and (b) be actually useful for analysing organizations and supporting decision making related to organizational purposes (this is DEMO added-value). Points (a) and (b) are detailed below.

### DEMO Models Quickly Provide a Picture of an Organization

*A simple picture of business processes with transactions and responsibilities.* Interviewees converge: DEMO is not only about describing a situation, but also changing the way people are looking at the organization: "It is constantly in the back of your mind when you are looking at things". Almost all interviewees spontaneously mentioned the words "responsibilities" and "transaction"; some also used the terms "act" and "fact". X1 explains: "Based on the transactions, coordination and production acts, you have precise definitions of authority, responsibility, competence and delegation." For X2, DEMO allows to "concentrate on the interface [between transactions], so the story became very simple."

*As many models as modellers?* Interviewees converge in saying that two DEMO models of a given situation designed by two modellers would usually not be identical. The cause is modellers' different ways of analysing things as being essential or not. Still, no interviewee suggested that those models would be inconsistent.

*DEMO has a good return on modelling effort (RoME).* All interviewees praised DEMO about its RoME. X3 declares: "RoME is one of the reasons why I liked very much DEMO, because it was very very efficient to get highly complicated matters clear." X10 thinks that without DEMO, people working on the project would have used more time to define roles.

### DEMO Supports Organization Analysis and Decision Making Related to Organizational Purposes

*Positive experiences with DEMO.* All interviewees would use DEMO again in their project in case they would have to do the project again. X11 is enthusiastic: "I think it is marvellous. I work smarter and not harder as a DEMO added value and focus on things that are worth managing." Many interviewees use the phrase "it works" to express how practical they thought DEMO was, e.g.: "so I saw that it worked and how fast it worked." X4 explains the success of his project: "For me DEMO was the key and I identified at least 2 or 3 elements that I am 90% sure that we would not have done without it."

*DEMO added value depends on project goals.* When being asked about DEMO added value, all interviewees were prolix. Here are the main points that interviewees mentioned. They are not independent from each other. DEMO helps in defining responsibilities without assigning them: "first, define; then, assign". The construction model is very often mentioned as being "the most outstanding benefit of DEMO". "If you don't use DEMO (…), it will be a complete different picture with all kinds of roles that are nearly close to the actual way of working" (X11). "DEMO enables you to pinpoint what is exactly happening and […] also makes sure that your model is consistent" (X4). X5 explains that, together with the existence of the construction model, the consistency between all DEMO models definitions is the source of DEMO added value. Indeed, it ensures the completeness and the consistency of the DEMO models produced on a project. About completeness, X4 says: "DEMO brings out new facts. You see, after 4 years, this fact (…) had not been identified in the hundreds of meetings and all the ARIS drawings they had." For X1, "DEMO is computable. It means that I can handle large organizations as easily as small. DEMO gives the capability of systematically deriving the map of authorities and responsibilities that have to be fulfilled to generate the acts that are associated with this. This is the basis for human department organizational design." X11 says that DEMO "brings you to the core of your business" and "allows distinguishing between what you have to manage and what you don't have to manage." Modelling with DEMO "brings clarity", "helps you get rid of non-relevant things". Interviewees are very positive about DEMO models being "concise". A DEMO founder claims that "DEMO brings proper knowledge to support decisions to change an organization".

*Tentative explanation of DEMO users' satisfaction and DEMO added value.* Due to many quotes from the interviewees, we think that DEMO added value and interviewees' positive experiences with DEMO may be explained by two main factors: DEMO's RoME and DEMO's fidelity with real world phenomena. For the latter point, various interviewees express the fact that DEMO reflects their real world: "DEMO is an abstraction; you can fit the real things in it" (X7). X3 says that DEMO allows to model "what is actually going on, (…) all the tricks, the non-official way, the way things really work, to get them on the table whereas most modelling is done based on procedure manuals but it is not the way it works." Many interviewees also spontaneously state that DEMO models are stable in the time. X6 adds: "And they are very stable, but they are easy to expand, to change." We explain this "fidelity with real world phenomena" by the theories (Habermas' and psi-theories) on which DEMO relies: "Jan Dietz [a DEMO founder] put in the middle of his model the human, the human who can decide. He put the human role as a main factor, at the centre." (X11)

## Conditions for Successfully Applying DEMO in a Project

*Management strongly supports the use of DEMO and the project.* If usual conditions of success of projects have their importance, the support from the management is especially seen as "crucial" when working with DEMO. We explain it because of DEMO bringing transparency about how the organization works (there might be people who do not want it), because this transparency may lead to decisions towards changes in the organization (those decisions are subject of resistance to change) and because resistance to change can also occur towards DEMO unusual way of thinking.

*Management wants transparency.* A DEMO specific condition is that management has to want transparency [2]. X5 explains: "DEMO makes things totally transparent, and in some cultures that is not what they wish." X3 suggests that DEMO may not be applied in organizations where people have a power-oriented mind-set. Besides, DEMO modellers should have access to people who are knowledgeable about how the organization works.

*A DEMO expert works on the DEMO modelling project.* Interviewees converge about this point: "training and advice from an experienced person is a prerequisite." Besides, people working with this expert "must be very aware of the conceptual basis of DEMO [...]. It should wise that they have followed a course on that." (X2)

### 4.7 DEMO Ease of Use – Convergence and Divergence

For our interviewees, DEMO is easy to use when you know it well. X1 said: "Because of my training with DEMO, that way of reasoning is implicit in my mind".

**Skills and Competencies to Model with DEMO – Convergence**
Interviewees converge in saying that, as DEMO is an abstraction, a certain level of abstraction capacity is useful. Besides, having experience with enterprise organization is an asset, whether this experience is in one organization or in several ones as for external DEMO consultants. For interviewers, it is primordial modellers are "well trained in DEMO", "DEMO professionals"; and that people working with them "understand some of the principle foundations of the approach". Working with DEMO also requires rigor, preciseness. The soft skills that interviewees mention are social capabilities, "open-mindedness" and communication capabilities, analysis capacities "to filter out what is really happening" and being to rephrase the models in natural language according to the capacity of abstraction of their interlocutors, namely when "presenting the models to the business".

**Risks of Mistakes When Using DEMO and How to Mitigate Them – Convergence**
Modelling risks with DEMO are: to produce a model that is not complete if you forget a transaction (a DEMO founder), to identify a transaction that is not essential, to identify something as a transaction whereas the thing is not a transaction (X6) or

whereas the thing is part of another transaction (X8) and to employ the theory in the wrong way when you think you have understood it and you actually have not (X2). For a DEMO founder, DEMO trainings teach to avoid these traps. But for X8, it is not only a matter of training, but also "a matter of doing. If you say for instance that "proposal" is a transaction, then I would say no. I would say it is only a request and promise. And if you say "proposal" is an end result, then you are building a system for proposals. (...). That is wrong with huge consequences."

**(Dis)ease of Learning – Divergence**
X2's following quote sums up well interviewees' points of view: "my experience is that some people understand it in a few hours and some people never understand it." X3 shares his experience: "We explained the model to the executive level people, who very quickly understood it, because they understood the decision making responsibility concept which is native to the construction model." Two DEMO consultants admitted to lead business process analysis workshops by applying DEMO way of thinking without mentioning it, "not to bother people". A DEMO founder says: "I would say, from my experience in teaching it, that it is not really difficult if I think of people who are able to abstract and have sufficient experience in organizations and know the world." Still, an interviewee, otherwise enthusiastic about DEMO, admitted: "The way how Jan Dietz [a DEMO founder] brought DEMO to VISI was very complicated. (...) DEMO way of thinking is complicated. Then you have a very small group of people who can use DEMO." For another person: "there are parts of DEMO that are really not easy to understand." Another one, although knowing DEMO well, mentions an "obscure terminology".

**4.8 DEMO Models Usage: Not All Models Are Used Each Time – Convergence**

If some interviewees would advise to use all types of DEMO models, most of them express that they "have been using parts of DEMO", e.g. the construction model is always produced, the fact and process model sometimes. A DEMO founder says: "If you only want to talk about the organization in the sense of assignment of people or organizational functions to actor roles, then it is most of the time sufficient to have the construction model, often combined with the process model. (...) The action and state model only really are necessary if you are going to develop or select applications."

# 5 Summary

*Research bias.* This paper mainly reflects the views of highly educated Dutch people with an engineering or information systems background. None of them is from the "Y generation", who is supposed to learn and think a bit differently. Besides, for several interviewees, 4 to 8 years elapsed since they have been in contact with DEMO for the last time; their projects' success (§4.1) may also give a positive flavour to everything related to them, especially years after. Furthermore, interviewees who are DEMO consultants may have an interest in praising DEMO. Interviewer's profiles and their degree of knowledge of DEMO may also impact the study, even if a well-defined research method is supposed to mitigate this risk. Interviewees mainly referred to two projects during the interviews (Air France KML Cargo IT merge and VISI). Still, some interviewees had experience with several DEMO projects and their views reflect their overall experience with DEMO. The findings of this qualitative analysis are restricted to the contexts mentioned by the interviewees and would require further investigation so that we could generalize them to any DEMO project.

*Research contribution about DEMO.* As most of our research participants are DEMO professionals or have been acquainted with DEMO for a long time, we expect that their views can provide a fair picture of how the use of DEMO in practice is actually seen by DEMO experienced field people. Besides, in many cases, the investigation of apparent interviewees' divergences showed that these divergences often come from a difference of interpretation of some words – e.g. "method", or from the scope of interviewees' answers – e.g. for DEMO being descriptive or prescriptive. Having disclaimed these apparent divergences, we may say that interviewees' views usually converge or complete each other's. Still, some divergences actually appeared, namely about the way DEMO is taught to its newcomers. Interviewees are very positive about DEMO being effective in fulfilling its purposes and these purposes seem to be relevant to the business. This can be related to how Aier and Fischer define the "usefulness of an artefact" [13] and the "utility of a design theory" [7] p158: "the artefact's ability to fulfil its purpose if the purpose itself is useful. The purpose of an artefact is only useful if it is relevant for business." Interviewees praise DEMO's RoME. Still, strong conditions are required so that DEMO can be effective: learning DEMO requires a strong investment, not every organization is ready for transparency and every project does not benefit from the support of the management. Besides, current DEMO tooling and DEMO way of working would require improvements.

*Conclusion.* During interviews performance and analysis, we experienced that Aier and Fischer's criteria [7] are interdependent when used to observe an artefact – e.g. the perception of an artefact's ease of use depends on the artefact's user profile and, in turn, this profile notion refers to the scope of the artefact's scope of application. Interviews seem to be a relevant means to gather research material about DEMO in practice. Further work should be performed to determine whether it would be the case with other artefacts and under which conditions. We do not know either the importance of the information about the use of an artefact in practice that could not be collected with interviews.

**Acknowledgments.** We warmly thank our interviewees, amongst whom are Jan Dietz, José Tribolet, Henk Schaap, Hans Zwitzer, Michael Kimman, Martin Op 't Land, Paul Ensink, Hans Jongedijk, Hans Mulder, Cees Buijs, Anton De Vroomen and Jos Hamilton.

## References

1. Dietz, J.L.G.: Enterprise Ontology - Theory and Methodology - Outline of the book. Springer (2006)
2. Op 't Land, M., Zwitzer, H., Ensink, P., Lebel, Q.: Towards a Fast Enterprise Ontology Based Method for Post Merger Integration. In: Shin, S.Y., Ossowski, S. (eds.) Proceedings of the 24th Annual ACM Symposium on Applied Computing (SAC), Honolulu, Hawaii, USA, pp. 245–252 (2009)
3. Dias, D.G., Mendes, C., Mira da Silva, M.: Using Enterprise Ontology for Improving the National Health System-Demonstrated in the Case of a Pharmacy and an Emergency Department. In: Filipe, J., Dietz, J.L.G. (eds.) Proceedings of the International Conference on Knowledge Engineering and Ontology Development (KEOD), pp. 441–451. SciTePress, Barcelona (2012)
4. Décosse, C., Molnar, W.A., Proper, H.A.: A Qualitative Research Approach to Obtain Insight in Business Process Modelling Methods in Practice. In: Grabis, J., Kirikova, M. (eds.) PoEM 2013. LNBIP, vol. 165, pp. 161–175. Springer, Heidelberg (2013)
5. Miles, M.B., Huberman, A.M., Saldaña, J.: Qualitative Data Analysis: a Methods Sourcebook. SAGE Publications, Inc. (2013)
6. Walsham, G.: Interpreting Information Systems in Organizations. John Wiley & Sons, Inc. (1993)
7. Aier, S., Fischer, C.: Criteria of Progress for Information Systems Design theories. Inf. Syst. E-bus. Manag. 9, 133–172 (2010)
8. March, S.T., Smith, G.F.G.: Design and Natural Science Research on Information Technology. Decis. Support Syst. 15, 251–266 (1995)
9. Myers, M.D., Newman, M.: The Qualitative Interview in Information Systems Research: Examining the Craft. Inf. Organ. 17, 2–26 (2007)
10. Kaufmann, J.-C.: L'entretien compréhensif. Armand Colin (2011)
11. Yin, R.K.: Case Study Research: Design and Methods (Applied Social Research Methods). SAGE Publications, Inc. (2008)
12. Seligmann, P.S., Wijers, G.M., Sol, H.G.: Analyzing the Structure of Information Systems Methodologies, an Alternative Approach. In: Maes, R. (ed.) Proceedings of the First Dutch Conference on Information Systems, Amersfoort, The Netherlands, pp. 1–28 (1989)
13. Aier, S., Fischer, C.: Scientific Progress of Design Research Artefacts. In: Proceedings of the 17th European Conference on Information Systems, ECIS (2009)

# A Pension System Redesign Case - Limitations and Improvements on DEMO

Akiyoshi Araki and Junichi Iijima

Tokyo Institute of Technology
2-12-1, Ookayama, Meguro-ku, Tokyo, Japan
{araki.a.aa,iijima.j.aa}@m.titech.ac.jp

**Abstract.** There are many methodologies and methods proposed for modeling Business Processes. Each of them has advantages and disadvantages. In this research we focus on Business Process Redesign in order to reveal limitations on the current syntax of DEMO and also its way of working. DEMO has been proven and recognized as an effective modeling method for business process at design level. However, the effectiveness of DEMO in the case of redesign of business process is not sufficiently documented.

In this study, we applied DEMO to model the Industry Pension System utilizing the National Identification Number in Japan in order to investigate the effectiveness of applying DEMO in redesigning. We discuss the effectiveness and limitations of DEMO in redesign based on the Construction and Process Models. Consequently, we propose additional constructs to be added to DEMO's syntax, as well as some guidelines for DEMO's way of working in redesign situations. A new notation system for the improvement of DEMO is also discussed in our research.

**Keywords:** DEMO, Business Process Redesign, National Identification Number, Pension System.

## 1 Introduction

Business process redesign is sometimes required for a case of reorganization or an introduction of new system in an enterprise. Nowadays, several institutions are increasingly developing and using cloud services. This implies the need of redesigning in such institutions' processes. The redesign aims at helping organizations, which includes not only enterprises but also governments, to fundamentally rethink how they do their work in order to dramatically improve customer service, cut operational costs, and become world-class competitors [1]. The redesign process generally consists of (i) setting a goal for the redesign, (ii) reviewing and analyzing As-Is process, (iii) designing a To-Be model in a certain method and (iv) testing, evaluating and verifying the To-Be process.

The Design and Engineering Methodology for Organization (DEMO), which is developed by Dietz [2], has been proven as an effective method for modeling of business process at design level [2, 3]. DEMO method has been applied for several case studies. Joop [3] applied DEMO for designing the information organization, Marien and Martin [4] linked DEMO methodology and Normalized System approach, Sanetake et al. [5] investigated how they can reduce exceptions and cancellations in business processes based on DEMO. João et al. [6] pointed out some issues of state of art approaches and proposed a value oriented analysis. However, the effectiveness of DEMO for redesigning business processes has not been sufficiently documented.

This research focuses on the effectiveness of DEMO in the case of business process redesign and discusses some limitations of DEMO. A contribution of this paper is proposing new components for DEMO's way of working, its syntax and a corresponding new notation system. Our proposal is based in a case study of the redesign of a pension system in Japan.

This paper has the following structure: section 2 presents theoretical background of DEMO. Section 3 explains the details of Industry Pension System in Japan and the information sharing system using the National Identification Number in the future. Section 4 shows the case study of the redesign, namely how the business process of pension service will be changed by the introduction of National Identification Number. We use the construction and process models from DEMO. In section 5, we discuss why the proposal is required and how it better contributes for redesign initiatives. Final part of paper includes conclusions and some remaining limitations.

## 2 Theoretical Background of DEMO

DEMO is a methodology for the engineering and implementation of organizations [2]. It can reveal the essential structure of business process and simplify the structure of organization by an ontological model which describes the core of the organizations. DEMO is based on the Performance in Social Action theory, PSI- or $\Psi$-theory. The $\Psi$-theory consists of four axioms and one theorem, i.e. the operation axiom, the transaction axiom, the composition axiom, the distinction axiom and the organization theorem [2].

- The operation axiom states that the operation of the organization consists of the activities of actors who perform two kinds of acts; production acts (P-acts) and coordination acts (C-acts). By performing P-acts, the actors contribute to achieving the purpose or the mission of the enterprise. By performing C-acts, the actors enter into and comply with commitments towards each other regarding the performance of P-acts.
- The transaction axiom states that C-acts are performed in transactions that always involve initiator and executor. They aim to achieve a particular result, the P-fact.
- The composition axiom states that transactions are related to each other, i.e., a transaction is enclosed in another transaction, or a transaction is self-activated.
- The distinction axiom states that there are three distinct human abilities playing a role in the operation of actors, i.e., performa, informa and forma ability which relates to ontological action, infological action and documental action, respectively. Actors who use the performa ability to perform P-acts are called business actors (B-actors). The performa ability is the essential human ability for

doing business. Actors who use the informa and forma ability to perform P-acts are called intellectual actors (I-actors) and documental actors (D-actors), respectively.
- The organization theorem states that the organization of an enterprise is an integrated social system of B-organization, I-organization and D-organization.

Based on Ψ-theory, DEMO allows the specification of an ontological model based on four coherent and consistent aspect models: (1) the Construction Model (CM) specifies the construction of the organization system by the identified transaction kinds and the associated actor roles, as well as the information links between the actor roles and information banks; (2) the Process Model (PM) contains the specific transaction pattern of the transition kind for every type in the CM; (3) the Action Model (AM) specifies the imperatively formulated business rules that serve as guidelines for the actors in dealing with their agenda, and (4) the State Model (SM) specifies the state space and the transition space of the production world with object class, fact types, result types and ontological coexistence rules.

The Construction Model contains actor roles, transactions and system boundary. Each transaction involves two actor roles, except when an actor role fulfills both the initiator and executor roles in that transaction, the case of self-activated transactions. Normally, the system boundary is settled based on the scope of business process and divides the actor roles into internal and environmental actor roles, and the transaction types into internal and boundary transaction types. An internal actor role is an actor role that is executor of an internal transaction type or a boundary transaction type. Environmental actor roles are initiator or executor of a boundary transaction type. We propose a nuance and new terminology, where we consider external actor roles that initiate/execute external transaction types (i.e., nor internal nor boundary ones). External transaction types as well as the external actor roles are usually disregarded due to being outside the scope. This paper will provide an example of exception where external transactions are important to be represented.

## 3 Utilizing National Identification Number in the Pension Service

### 3.1 Pension Plans in Japan

A pension supports the basic life requirements of a retired citizen. In Japan a universal pension system was started in 1961. Nowadays, Japan has the highest rate of aging population in the world, over 40 % of citizens will be over 60 in 2050[7], therefore, the pension system becomes even more significant. The pension system in Japan has three layers [8]. The first is a basic pension of which the premium is constant and all persons above 20 years old have an obligation to affiliate. The second is a public pension including welfare pension insurance and mutual aid pension of which the premium is a certain ratio of the income and those who are salaried workers must affiliate. The retirement pensions of the basic pension and public pension are provided by Japan Pension Service (JPS). The third layer is a private pension including industrial pension and defined contribution pension plan. Industrial pension consists of several types of plan. For example an employee pension fund (EPF) plan, a defined benefit (DB) plan, company pension fund plan and so on. All of these are a complement of the basic and public pensions. The retirement pension of EPF plan and DB plan are provided by an EPF. This work focuses on the EPF and DB plans.

More details about the EPF and DB plans are described below. The pensions are provided to the retired persons when they are no longer earning a steady income from a company. Both the employer and employee are required to contribute money to an EPF during their employment for the EPF plan. The contribution from an employee is a part of a premium of welfare pension. EPF partly deputizes to invest for JPS and EPF can utilize a scale merit which allows EPF to make larger benefit. The benefit is added to the pension payment. An EPF associated with more than 4 others can form a Pension Fund Association (PFA). A PFA provides an integrated pension payment those who opt out of an EPF or change EPFs before reaching a certain period in an EPF.

A DB plan is a plan in which the benefit on retirement is determined by a formula based on the employee's earning history, tenure of office, age and so forth, so that the amount of the retirement pension is defined in advance. The company principally contributes to the DB plan and the member also can contribute for it. A DB plan can be of a funded or of an unfunded type. In a funded plan, contributions from the employer, and sometimes also from plan members, are invested in an EPF. In an unfunded plan, no assets are set aside and the contributions are invested by a trust company, insurance company or investment advisory firm contracted with the employer.

## 3.2 National Identification Number

A national identification number is used in many countries, e.g. REAL ID or Social Security number in the U.S., National Insurance Number in U.K. and Personal Identity Number in Sweden. The number is mainly used to check identity of residents for the purposes of social security, taxation, national health care and others. In most cases, the number consists of more than 9 digits and letters related to the date of birth or sexuality and it is issued at birth or when people reach a certain age. A government using a national identification number supports precise and efficient taxation, insurance and social security system to the government. It also supports one stop service to people.

The government of Japan has been discussing about the introduction of a national identification number since 1994. The first identification number called the Resident's Card Code has operated since 2002. However, the use of the code has been highly restricted due to privacy and security issues and no private enterprises have been able to use the code. Therefore few organizations can get benefits from the use of the Resident's Card Code. The government finally decided to introduce another identification number for enabling the management of personal information including name, address, income, tax, pension payments and so on. The new number will be issued in 2016 and used not only in government but also in private enterprises in the future. The application range of the number in the private domain will include social insurance service and taxation.

The introduction of the new ID will allow organizations to share the resident's information. Figure 1 shows a schematic diagram of a prospective information sharing system using an ID. The ID will be provided by local offices. Each information providing organization manages their own resident's personal information by the ID. Each organization also has a converting system from the ID to a Code for providing personal information for other organizations. For example, when organization A requests a specific personal information to organization C, the organization A sends the name of information and Code A. The Code A is converted to Code C in Information Service Network System, then the organization C receives the name of information and Code C. Finally, the organization C sends the corresponding information to organization A directly.

The information sharing system using ID also supports many convenient services for both individuals and administrative offices including one stop service and push services. Individuals will not need to gather many kinds of documents from different administrative offices for an application for insurance or allowance. Thus, administrative offices can cut the cost for the confirmation of documents and can prevent an artificial mistake and electronic data alteration or the like. Thus, the introduction of national identification number aims to reduce the work load of both residents and organizations.

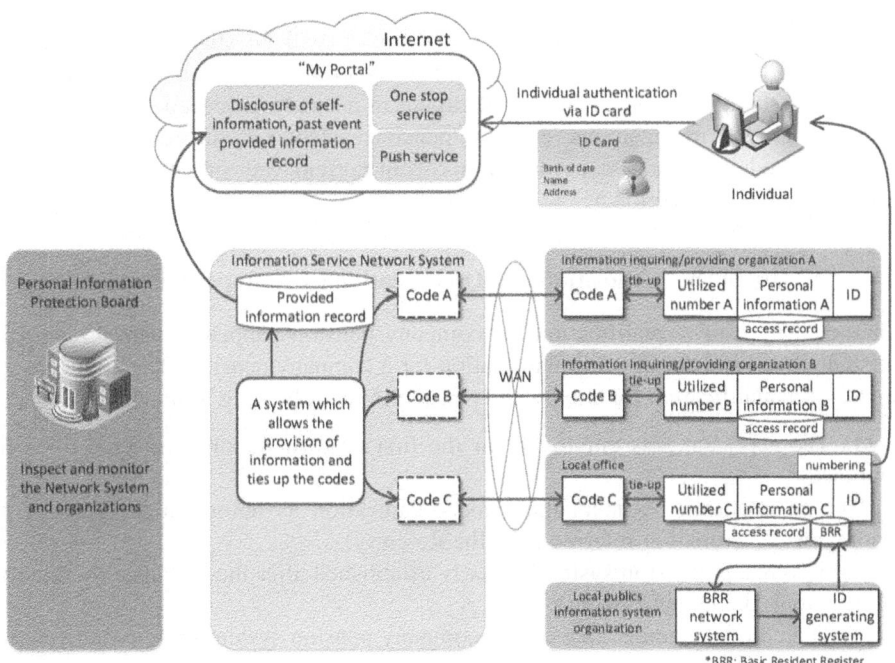

(Special Committee on Computerization Promotional Measures [11]: http://www.nga.gr.jp/news/25.4.18zyouhoupt1111111.pdf, translated by author)

**Fig. 1.** A schematic for the information sharing system using identification number

## 4 Building of As-Is and To-Be Models

In order to build the models based on DEMO, we discussed about the business process of pension service with the Institute of Strategic Solutions for Pension Management which is an institution aiming for the improvement of the pension service. The industrial pension plan was targeted which is one of a fundamental pension in Japan.

There are 36 business processes for the administration of EPF and DB in Japan. The business processes include an acquisition of a qualification, a decision of the amount of pension payment, a confirmation of current situation of a recipient, and so on. These business processes are expected to become simpler by the introduction of national identification number and information sharing system.

In this section, two business processes are picked as an example in order to investigate how the business model will be changed by the redesign; the first one is the acquisition of a qualification which happens most frequently in the business process of pension service; the second one is a confirmation of the current condition of a recipient of which the process is expected to be drastically changed by the redesign. Then As-Is models and To-Be models are specified for the business process before and after the redesign, respectively. The As-Is business models are specified based on the practical business processes. The To-Be business models are specified based on a perspective that resident's information will be shared by local office, company, EPF and JPS using the national identification number. The specified models are, as mentioned in section 1, drawn in a newly proposed way. Our scope here includes the organizations related to pension system, i.e., the company, EPF and JPS. The models include some deviations from current way of modeling and these will be discussed in section 5.

### 4.1 A Case of the Acquisition of a Qualification for Company Pension

The acquisition of a qualification for company pension happens when an applicant gets employed and applies for qualification for a company pension. The applicant can apply for a qualification for a company pension in the following four cases.

(1) The applicant gets employed for the first time (hereinafter referred to as the Entrance).

(2) The applicant re-enters the same organization as before in the case of re-employment (hereinafter referred to as the Re-entry).

(3) A company pension system is newly established after the Entrance (hereinafter referred to as the Establishment).

(4) The company enters an existing company pension system (hereinafter referred to as the Company Admission).

The Construction Model (CM) of the current business process for the acquisition of a qualification for company pension and corresponding Transaction Production Table are shown in Figure 2 and Table 1, respectively. This CM includes an external actor

A Pension System Redesign Case - Limitations and Improvements on DEMO    37

role outside of the boundary, named Local Office, and an external transaction related to Local Office, which are usually disregarded. This will be discussed in section 5.

First, the applicant (CA01) requests a resident's card (T1) to a local office (CA02). After the applicant gets his/hers resident's card, the applicant applies for the qualification for company pension (T2) to the company (CA03). For the case of (2) the Re-entry, the applicant must submit a subscription deed which is issued at the Entrance. After the application, registrations to an employee pension fund (EPF) (CA04) and Japan Pension Service (JPS) (CA05) must be performed (T3 and T4). EPF and JPS accept the registration and make an entry in the registration book, then, send a notice of decision in order to let the company know the completion of registration. Afterward, the application will be completed. For the case of the EPF plan, the company must check the notices of decision from EPF and JPS (T5) in order to verify that there is no difference in the records. Then, the company makes "a notification about article 128" for the result of the verification and sends it to the EPF. The reason why both T3 and T4 are required is that EPF and JPS are independent even though both pay the pension to affiliates. The Process Model (PM) for this case is shown in Figure 3. The completion of T2 must wait the acceptances of T3 and T4. T5 also must wait the state of T2.

The To-Be CM for the acquisition of a qualification for company pension is shown in Figure 4. The model contains an actor role CA01 which does not have any transaction with other actor roles in order to compare with As-Is model. Corresponding Transaction Production Table is same with Table 1 without T1 and R01. The local office provides personal information of the applicant for the company through the information network, therefore no transaction exists between the local

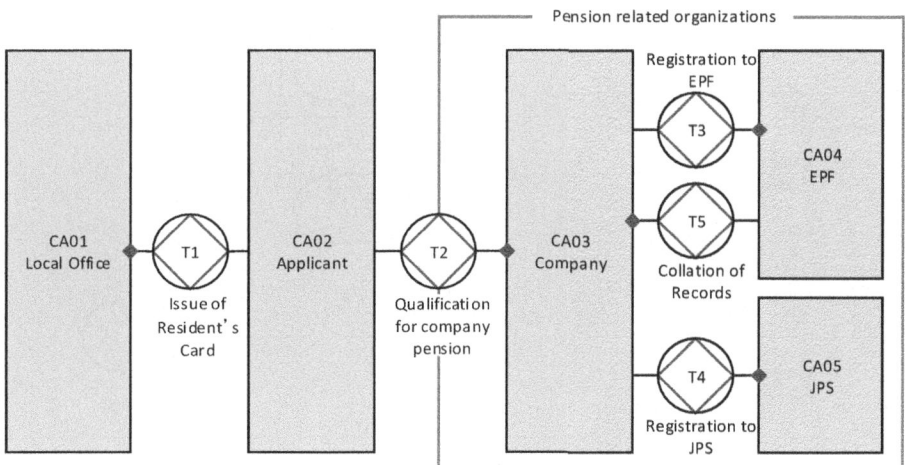

**Fig. 2.** As-Is construction model of the acquisition of a qualification for company pension

**Table 1.** Transaction Production Table corresponding to Fig. 2

| | Transactions | | Production |
|---|---|---|---|
| T1 | Issue of Resident's card | R01 | Resident's card has been issued |
| T2 | Qualification | R02 | the applicant has been qualified |
| T3 | Registration to EPF | R03 | Registration to EPF has been done |
| T4 | Registration to JPS | R04 | Registration to JPS has been done |
| T5 | Collation of records | R05 | The records has been collated and an article 128 has been made |

office and the applicant. T5 becomes a self-activated transaction because EPF can collate the records of JPS by itself through the network. A corresponding PM is shown as Figure 5. The PM also contains CA01 even though the actor has no longer any transaction with other actors in order to compare with Figure 2. The model also contains the fact banks in order to express which transactions use the information network. One can easily see that the model of the changed process is simpler than that of current process. In changed process, there is no waiting condition related with T5.

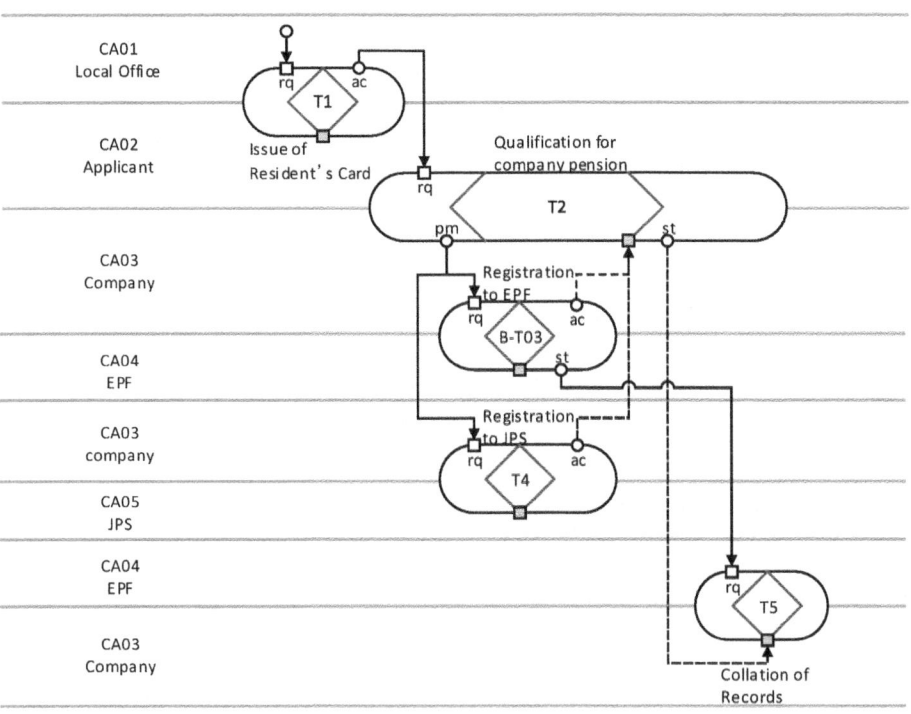

**Fig. 3.** As-Is process model of the acquisition of a qualification for company pension

A Pension System Redesign Case - Limitations and Improvements on DEMO    39

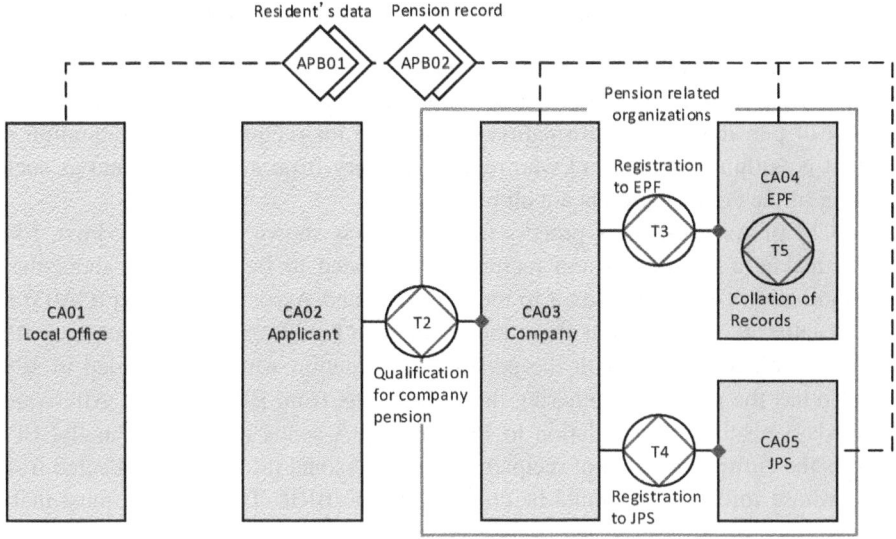

**Fig. 4.** To-Be construction model of the acquisition of a qualification for company pension

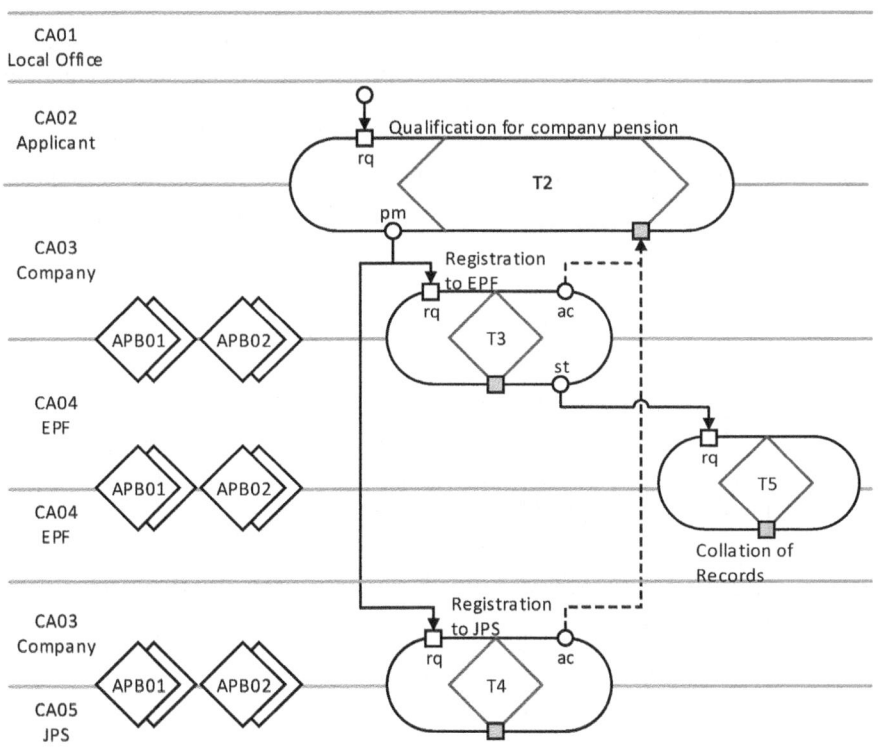

**Fig. 5.** To-Be process model of the acquisition of a qualification for company pension

## 4.2 A Case of the Confirmation of Current Condition of a Recipient

The other case, the confirmation of current condition of a recipient happens when EPF confirms the current condition of recipients of employee pension for precise payment of pension. This confirmation is not done for recipients whose duration as recipient is within one year and who receive a salary from a company, and as such, the target of the confirmation is not all recipients.

The CM of As-Is business process for this case is shown in Figure 6. First, EPF (CA02) makes a file of pension recipients who need to be confirmed about their current situation (T1) and sends it to the Pension Fund Association (PFA) (CA03) in order to gather the information about the recipients (T2). Then, the PFA requests JPS (CA04) to collate the file with the personal information which is recorded in JPS. PFA also has the choice of gathering the information from BRR directly. Afterward, JPS sends a result of the collation to PFA and PFA sends it to EPF. Finally, EPF confirms the current situation of recipients whose personal information collected from JPS or whose information cannot be collected in the BRR. The recipient must make proof that they are still alive and of their current address (T4). EPF also has a choice to do T4 directly, but that way requires a much bigger workload.

The CM of the To-Be business process is shown in Figure 7. As with the case of the acquisition of a qualification for company pension, the actor roles which no longer have any transaction are expressed in To-Be model. The CM changes dramatically and the model has only one transaction, i.e. EPF can confirm the current situation of recipient by itself through the information network and no transactions with other organizations are required. The drastic change of the model means the ID system works remarkably for this business process and it makes the process much simpler.

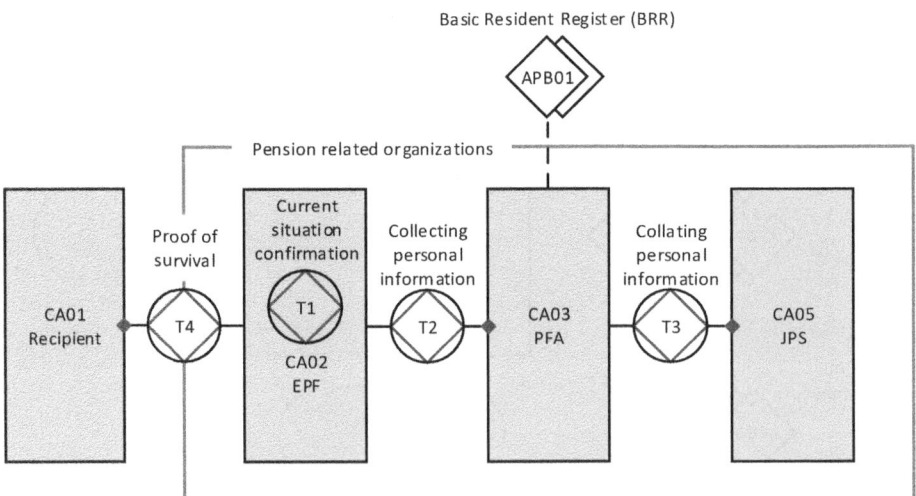

**Fig. 6.** As-Is construction model of the confirmation of current condition of a recipient

**Table 2.** Transaction Production Table corresponding to Fig. 6

|    | Transaction                    |     | Production                            |
|----|--------------------------------|-----|---------------------------------------|
| T1 | Current situation confirmation | R01 | Current situation have been confirmed |
| T2 | Collecting personal information| R02 | Personal information have been collected |
| T3 | Collating personal information | R03 | Personal information have been collated |
| T4 | Proof of survival              | R04 | The survival has been proved          |

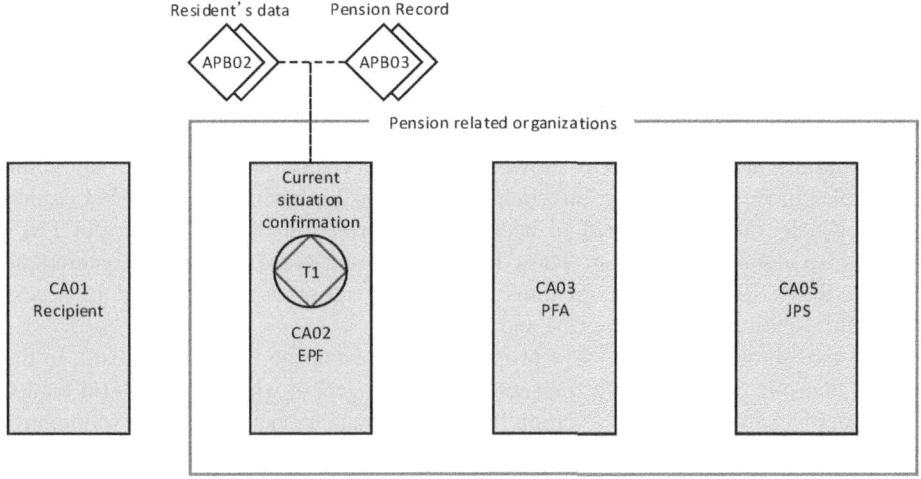

**Fig. 7.** To-Be construction model of the confirmation of current condition of a recipient

## 5 Discussion

We found that current DEMO'S way of working and syntax has several limitations for revealing the change of business model by the redesign through the case study. The objective here is to clarify the limitations of the current way of working and syntax, and to propose a new way of working and syntax for addressing the limitations. Models in section 4 deviate from current rules and already contain some of our proposals of improvement.

### 5.1 A Case of the Acquisition of a Qualification for Company Pension

The To-Be CM and PM deviate from the current way of working in order to clarify the effect of the business process redesign on each model. In the current way of working the external actor role and transaction are usually not expressed in the models because they are outside the scope. In this case, however, the redesign affects not only the internal transactions but also the external transactions like T1 as shown in Figure 2 to Figure 5. The extinction of the transaction between Applicant and Local

Office simply means that the workloads of applicant no longer exist after the redesign. This change is important in this case because the applicant must be cared the most in the pension service.

Moreover, CA01 no longer has any transactions with other actor roles in the To-Be models. Current DEMO way of working does not consider drawing such "alone" actor roles. However, without CA01, it is not clear that CA02 no longer has the external transaction which exist in As-Is model. Therefore, all the actor roles which exist in the model before the redesign should be contained in the model after the redesign too in order to investigate which transactions are changed or deleted by the redesign and which actor roles now become "alone".

In this research, the personal information like a resident's information and pension record shared by the organizations through the information network is expressed as fact banks in the PM. DEMO methodology will be more useful if the information bank is represented in PM too as shown in Figure 5 because the CM can reveal which actor roles have access to the information banks in the business process but cannot reveal which transactions need to access the information bank. The current PM's syntax also cannot represent it. Thus, DEMO's current syntax does not contemplate the possibility of expressing, in diagrams, which transaction steps need to access which information banks in the CM and the PM.

The way shown in figure 5 is a proposal to represent the information bank in the PM so that one can see which transactions will be access which information banks. For example, showing the information bank on a boundary between CA03 (company) and CA04 (EPF) indicates the information bank will be used for the transaction T3 when the EPF gets the personal information in order to register it to a record book. Now, it can be clarified which actor roles use which information banks, and which transactions access which information banks and when it happens in the process as represented in the PSD of the PM. This information about the timing of the usage of information bank will be useful for a next step of the redesign, namely, the simulation of the business processes. Although the PM based on current DEMO's syntax does contemplate the Information Use Table which represents such information, the newly proposed notation is indeed more effective because it links directly the transaction and information bank and is more intuitive than looking at the Information Use Table. In fact, there is a business process which requires many notations about which transactions require which information banks, as a result the model is confusing. The model will be simplified by applying the proposed way of notation. However, the question raises: how to represent in diagrams which process steps are linked with which information banks? A new notation is required to answer this question.

We attempted addressing this question by proposing a new notation system in the process model as shown in Figure 8. In figure 8, now, each information bank is linked with the transaction steps with green dashed line. In DEMO methodology green color illustrates infological issues, so it is natural that we use the green color to represent these information access links. This figure shows more intuitively which transaction steps use which information banks.

Now, a problem still remains. Considering the change of business process, the workload of each actor role should decrease after the introduction of the information sharing system. For instance, the applicant does not have to submit documents of personal information for T2 any longer because the company can get such information from the information network in the To-Be business process. However, the change of workloads cannot be shown in neither the CM nor the PM in current way of modeling because both models cannot contain a quantitative information such as amount of documents, time and information transfer required for each transaction. Putting quantitative information into the models requires more discussions and case studies. However, there are several solutions for this e.g. a simulation on which we can add weight of the workloads to the model or an advanced GSDP [6].

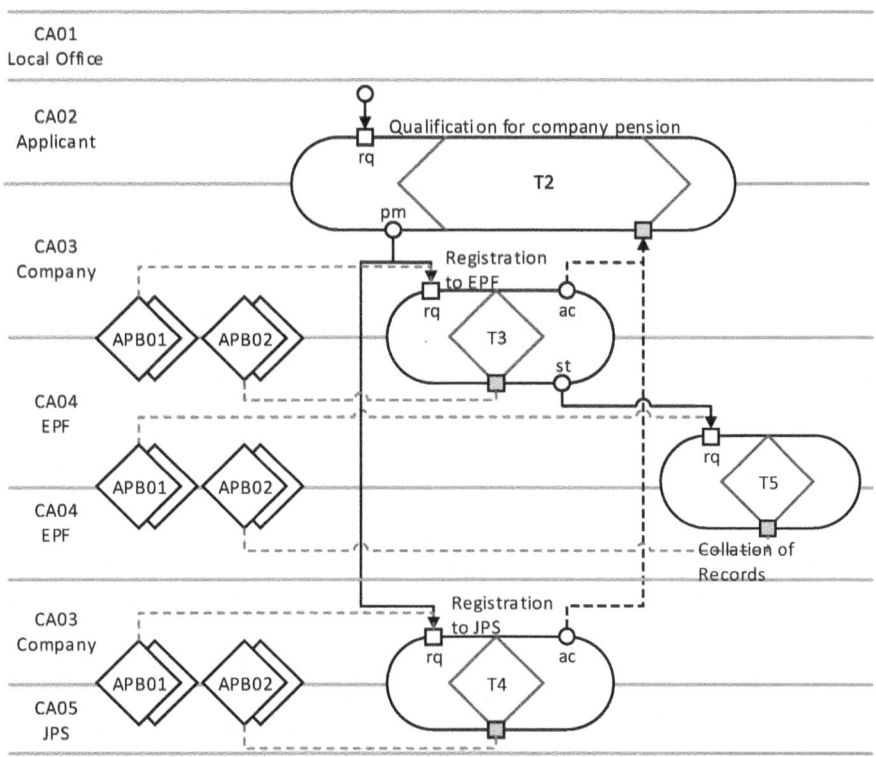

**Fig. 8.** Newly proposed notation for the process model

### 5.2 A Case of the Confirmation of Current Condition of Recipients

This case study shows that DEMO is also very effective for a redesign. DEMO can obviously reveal that the business process will be very simplified by the introduction of information sharing system. Comparing As-Is and To-Be CM, one can easily see that the actor CA01 (applicant) becomes an external from an environmental actor role required by the redesign, which means that no external actor roles will be needed after the redesign and this process will be instantly done. As with the previous case, the proposed way of notation reveals that CA01, CA03 and CA04 no longer have any transaction with other transactions.

# 6 Conclusions

For the case of business process redesign, comparing As-Is and To-Be models is necessary to verify the effectiveness and practicality of the To-Be model. As discussed through the case studies about the business process redesign of the pension system. However, the syntax of current DEMO and also its current way of working have some limitations in revealing how the business process will be changed by the redesign.

Current DEMO's way of working does not consider external actor roles and transactions or actor roles which do not have any transactions even if those are related to important environmental actors. A proposal for addressing this problem is to express those actor roles and transactions even though the way deviates from current practice. This proposal is useful when comparing As-Is and To-Be models and investigating how the business process redesign affects the business process itself as well as involved actor roles

Current DEMO's syntax also cannot express the role of information bank in the PM. A proposal for a new way of representation of the information bank is shown above. The proposed way successfully expresses which transactions are linked with the information bank. This may help estimate how the information banks can simplify the business process in PM level and will make DEMO more useful methodology. Furthermore, the information banks are intuitively linked with the transaction steps in discussion part. Applying this new way for practical business process and investigating the effectiveness will be future work.

The proposed way of notation can also clearly show which actors are active or inactive and which transactions will be reduced by the redesign in some business processes.

Some limitations still remain in DEMO methodology. For example, the workloads cannot be expressed both in construction model and process model. Though there are several methods to address the problem including a simulation and advanced GSDP, putting the workloads into DEMO model needs more discussions and case studies.

**Acknowledgement.** The authors thank the research member of the Institute of Strategic Solutions for Pension Management for the discussion of the business process in the industry pension service.

# References

1. Dodaro, G.L., Crowley, B.P.: Business Process Reengineering Assessment Guide (1997)
2. Dietz, J.L.G.: Enterprise Ontology – theory and methodology. Springer, Heidelberg (2006)

3. de Jong, J.: Designing the Information Organization from Ontological Perspective. In: Albani, A., Dietz, J.L.G., Verelst, J. (eds.) EEWC 2011. LNBIP, vol. 79, pp. 1–15. Springer, Heidelberg (2011)
4. Krouwel, M.R., Op 't Land, M.: Combining DEMO and Normalized Systems for Developing Agile Enterprise Information Systems. In: Albani, A., Dietz, J.L.G., Verelst, J. (eds.) EEWC 2011. Lecture Notes in Business Information Processing, vol. 79, pp. 31–45. Springer, Heidelberg (2011)
5. Nagayoshi, S., Liu, Y., Iijima, J.: A Study of the Patterns for Reducing Exceptions and Improving Business Process Flexibility. In: Albani, A., Aveiro, D., Barjis, J. (eds.) EEWC 2012. LNBIP, vol. 110, pp. 61–76. Springer, Heidelberg (2012)
6. Pombinho, J., Aveiro, D., Tribolet, J.: Value-Oriented Solution Development Process: Uncovering the Rationale behind Organization Components. In: Proper, H.A., Aveiro, D., Gaaloul, K. (eds.) EEWC 2013. LNBIP, vol. 146, pp. 1–16. Springer, Heidelberg (2013)
7. World Population Prospects: The 2012 Revision. United Nations, New York (2012), http://esa.un.org/wpp/Documentation/pdf/WPP2012_%20KEY%20FINDINGS.pdf
8. Japan Pension Service, http://www.nenkin.go.jp/n/www/english/index.jsp
9. Special Committee on Computerization Promotional Measures, http://www.nga.gr.jp/news/25.4.18zyouhoupt1111111.pdf

# A New Action Rule Syntax for DEmo MOdels Based Automatic worKflow procEss geneRation (DEMOBAKER)

Carlos Figueira[1] and David Aveiro[1,2,3]

[1] Exact Sciences and Engineering Centre, University of Madeira,
Caminho da Penteada 9000-105 Funchal, Portugal
[2] Center for Organizational Design and Engineering,
INESC-INOV Rua Alves Redol 9, 1000-029 Lisboa, Portugal
[3] Madeira Interactive Technologies Institute,
Caminho da Penteada 9020-105 Funchal, Portugal
carlos.figueira.oelabuma@gmail.com, daveiro@uma.pt

**Abstract.** The current way of specifying Action Rules in the Design and Engineering Methodology for Organizations (DEMO) is ambiguous and leads to incomplete specifications that do not contain enough ontological information so that we can more systematically convert DEMO models to comprehensive Business Process Management and Notation (BPMN). With our proposal we now can specify – still at an ontological level – much more needed details and essential information for a more complete and close to automatic generation of BPMN models. Action rules are also the perfect spot to already specify functional and implementation requirements for the information systems supporting the Workflow Management System running such BPMN models. Thus we also contribute to bridge the huge gap between DEMO models and important implementation issues that arise at design time and should immediately be specified together with ontological elements.

**Keywords:** enterprise engineering, BPMN, BPM, DEMO, meta model, action model, action rules, syntax, workflow, information systems, requirements.

## 1 Introduction

Most IT projects fail to meet final user's expectations. From [1], where some case studies were made, a recent survey with 800 IT managers [2] [3], found that 63% of software development projects failed, 49% suffered budget overruns, 47% had higher than expected maintenance costs and 41% failed to deliver the expected business value and user's expectations. From these case studies, it was found that some of the common causes of software failures are: the lack of clear, well-thought-out goals and specifications, poor management and poor communication among costumers, designers and programmers [4], unrealistically low budget requests, and underestimates of time requirements, use of new technologies maybe for which the software developers don't have adequate experience and expertise and refusal to recognize or admit that a project is in trouble [1]. DEMO [5] is a renowned enterprise engineering method associated with a sound set of theories aiming to contribute to solve these serious problems. However many open ends exist. For example, the produced models are used mostly for isolated efforts of organizational analysis and providing support for discussing changes initiatives. And one of its main aspect models – the Action Model – is barely used in practice [6], although the founder of DEMO himself says it's probably the most important model and where all essential model information can be found and all the other 3 aspect models can be derived from [5].

Research presented in this paper is integrated in a wider research project aiming to develop a software platform to support collaborative and semantic web based production of organizational models and diagrams that specify organizational processes, human and software responsibilities, information flows, procedures and other kinds of organizational artifacts. Such models should consist in a continuously and collaboratively updated "picture" of the reality of an enterprise that guides its collaborators in: (1) the perception of the global "organizational self" [7] (2) the execution of their operational work and (3) the creative process of changing the organization itself [8], including or not the change or implementation of software and related information technology. Other widespread approaches such as ArchiMate [9] and BPMN [10] suffer highly from the lacking of a solid formal theory behind them and from ambiguous semantics [11] [12]. Our DEMO based approach, grounded in solid theory, aims to allow the generation of models that capture vital information of organizational responsibilities and information flows, normally neglected in other approaches. From these models – that have a very high-level of abstraction and are easy to share and comprehend – we aim to systematically derive increasingly detailed models down to runnable workflows and program code or manual work instructions. All models and all model artifacts are to be formally connected with each other in a coherent and semantically strong way, bringing immense power to our approach. Our prototype software platform is inspired on the Universal Enterprise Adaptive Object Model (UEAOM) [13] and is supported by the Semantic MediaWiki (SMW) software, to allow the integrated management and adaptation of: (1) enterprise models, (2) their representations, (3) their underlying meta-models, i.e., their abstract syntax, (4) the representation rules, i.e., the concrete syntax for the respective models, and (5) automated or semi-automated generation of: (i) runnable workflows; (ii) formal requirements for software to support such workflow and (iii) program code. All this for different modelling languages and also different versions of these languages, with an initial focus on DEMO and BPMN the most widespread standard used for workflow management systems. One of the components of our software prototype is called DEMOBAKER, having as a main goal to provide an efficient and standardized way of converting DEMO models into totally compliant BPMN models having clear semantics, an issue lacking in traditional BPMN approaches. In this paper we present an important step in this direction. Namely, we propose an improved meta-model for DEMO's Action Model also presented in the form of a new Action Rule Syntax. We focus on the problem that the current way of specifying Action Rules in DEMO is ambiguous and leads to incomplete specifications that do not contain enough ontological information so that we can convert DEMO models to comprehensive BPMN models. With our proposal we now can specify – still at an ontological level – much more needed details and essential information for a more complete and close to automatic generation of BPMN models. Action rules are also the perfect spot to already specify functional and implementation requirements for the information systems supporting the Workflow Management System running such BPMN models. Thus we also contribute to bridge the huge gap between DEMO models and important implementation issues that arise at design time and should immediately be specified together with ontological elements. We use the EU-rent case taken from [14] to exemplify and validate our contribution.

## 2 Research Method

According to A. R. Hevner [15] [16], Design Science Research – the Information Systems Research paradigm that we adopt – should be seen as a group of three closely related cycles of activities. These activities are depicted on Figure 1. Hevner claims that the individual application of these three activities in an isolated way does not constitute good design science research. Only the conjunction of the three can actually render good design science research with a valid output. In our research, and regarding the relevance cycle depicted on Figure 1, we identified a clear problem of ambiguity and lack of concise and essential information on current DEMO's action rule syntax. So an opportunity to devise a more sound and comprehensive syntax was at hand. Regarding the Rigor cycle, our research was supported by all the theoretical foundations grounding DEMO as well as the UEAOM patterns. The most important cycle is the Design cycle itself, out of which resulted our proposal of a new meta-model for DEMO's Action Model. An exhaustive and thorough evaluation was done with many iterations of this cycle where we would be adding new elements to DEMO's Action Meta Model and instantiating our new syntax with the EU-rent case and evaluating if it allowed to specify maximum ontological information in a concise and comprehensive way, normally not the case in DEMO's current standard Action Models. While instantiating and increasing the complexity of EU-rent case's action rules to a more realistic level, some times we found some concept in the meta-model should be unary, some other times other concepts should be binary, and, at other times, we found we would need to specify new concepts at meta-level like: *atomic action* and *flow*. So the proposal presented in this paper is the result of a long and thorough process of conceptual evolution and comprehensive instantiation, thus following the tenets of Design Science Research.

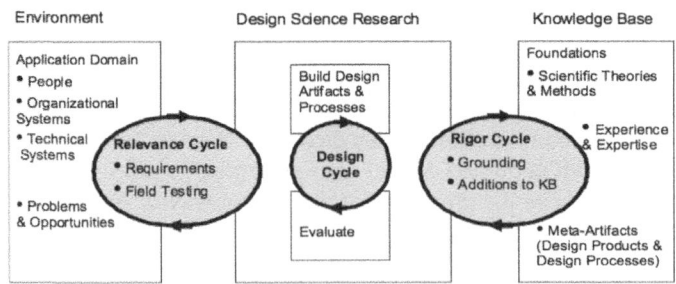

**Fig. 1.** Design science research cycles [17]

## 3 Background and Theoretical Foundations

### 3.1 DEMO's Operation, Transaction and Distinction Axioms

In the Ψ-theory [17] – on which DEMO is based – the operation axiom [5] states that, in organizations, subjects perform two kinds of acts: production acts that have an effect in the production world or P-world and coordination acts that have an effect on the coordination world or C-world. Subjects are actors performing an actor role responsible for the execution of these acts. At any moment, these worlds are in a particular state specified by the C-facts and P-facts respectively occurred until that moment in time. When active, actors take the current state of the P-world and the C-world into account. C-facts serve as agenda for actors, which they constantly try to deal with. In other words, actors interact by means of creating and dealing with C-facts. This interaction between the actors and the worlds is illustrated in Figure 2. It depicts the operational principle of organizations where actors are committed to deal adequately with their agenda. The production acts contribute towards the organization's objectives by bringing about or delivering products and/or services to the organization's environment and coordination acts are the way actors enter into and comply with commitments towards achieving a certain production fact [18].

According to the Ψ-theory's transaction axiom the coordination acts follow a certain path along a generic universal pattern called transaction [5]. The transaction pattern has three phases: (1) the order phase, were the initiating actor role of the transaction expresses his wishes in the shape of a request, and the executing actor role promises to produce the desired result; (2) the execution phase where the executing actor role produces in fact the desired result; and (3) the result phase, where the executing actor role states the produced result and the initiating actor role accepts that result, thus effectively concluding the transaction. This sequence is known as the basic transaction pattern, illustrated in Figure 3, and only considers the "happy case" where everything happens according to the expected outcomes. All these five mandatory steps must happen so that a new production fact is realized. In [18] we find the universal transaction pattern that also considers many other coordination acts, including cancellations and rejections that may happen at every step of the "happy path".

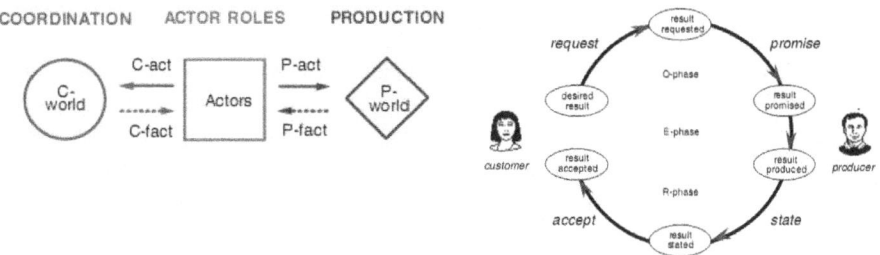

**Fig. 2.** Actor's Interaction with Production and Coordination Worlds [5]

**Fig. 3.** Basic Transaction Pattern [5]

Even though all transactions go through the four – social commitment – coordination acts of request, promise, state and accept, these may be performed tacitly, i.e. without any kind of explicit communication happening. This may happen due to the traditional "no news is good news" rule or pure forgetfulness which can lead to severe business breakdown. Thus the importance of always considering the full transaction pattern when designing organizations. Transaction steps are the responsibility of two specific actor roles. The initiating actor role is responsible for the request and accept steps and the executing actor role is responsible for the promise, execution and state steps. These steps may not be performed by the responsible actor as the respective subjects, may delegate on another subject one or more of the transaction steps under their responsibility, although they remain ultimately responsible for such actions [18].

The distinction axiom from the Ψ-theory states that three human abilities play a significant role in an organization's operation: (1) the *forma* ability that concerns datalogical actions; (2) the *informa* that concerns infological actions; and (3) the performa that concerns ontological actions [5]. Regarding coordination acts, the performa ability may be considered the essential human ability for doing any kind of business as it concerns being able to engage into commitments either as a performer or as an addressee of a coordination act [18]. When it comes to production, the performa ability concerns the business actors. Those are the actors who perform production acts like deciding or judging or producing new and original (non derivable) things, thus realizing the organization's production facts. The informa ability on the other hand concerns the intellectual actors, the ones who perform infological acts like deriving or computing already existing facts. And finally the forma ability concerns the datalogical actors, the ones who perform datalogical acts like gathering, distributing or storing documents and or data. The organization theorem states that actors in each of these abilities form three kinds of systems whereas the D-organization supports the I-organization with datalogical services and the I-organization supports the B-organization (from Business=Ontological) with informational services [19].

## 3.2 Business Process Management (BPM) and Its Notation – BPMN

As stated in [20], "Business Process Management (BPM) is the discipline that describes structured methods and techniques used to make a business process more efficient adaptive and effective for accomplishing a specific task within an organization" BPM techniques and methods also allow the identification and modification of existing processes in order to align them to future possibilities of change. BPMN, stands for Business Process Model and Notation and consists in a method for graphically representing the steps of a business process similar to a flowchart approach. The BPMN notation was specifically designed to allow the specification of the coordination of the sequence of organizational processes and the way that messages flow between activities, processes and participants.

## 3.3 Universal Enterprise Adaptive Object Model

The Universal Enterprise Adaptive Object Model (UEAOM) is a recent proposal from [13], consisting in a conceptual schema inspired in the the Adaptive Object Model (AOM) [21], a software architecture pattern for systems in which classes, attributes, relationships and behaviors of applications are represented as metadata, allowing them to be changed in runtime environment.

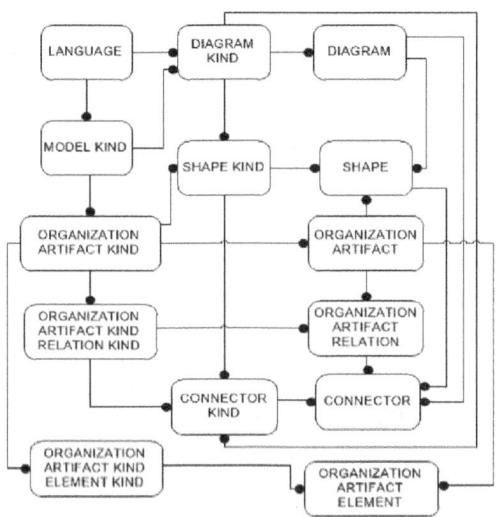

**Fig. 4.** UEAOM - simplified version

Figure 4 presents a simplified version of the original UEAOM, with the core classes only. A brief explanation and exemplification of the core classes relevant for the contributions of this paper follows. LANGUAGE – used to specify which languages are permitted in the diagram editor. Example: «DEMO v3.5». MODEL KIND – each language can have multiple model kinds, used to specify which kinds of models are permitted for a certain language in the Diagram Editor. Example: «Construction Model v3.5». DIAGRAM KIND – each model can have multiple Diagram kinds, used to specify which kind of Diagrams are permitted in the Diagram Editor for a certain model kind. Example: The Actor Transaction Diagram «ATD v3.5». ORGANIZATION ARTIFACT KIND (OAK) – used to specify which kinds of organization artifacts can be used in models. Example: «ELEMENTARY ACTOR ROLE v3.5». ORGANIZATION ARTIFACT KIND RELATION KIND (OAKRK) – used to specify which kinds of relations are permitted between OAKs. Example: «TRANSACTION KIND.executed by.ELEMENTARY ACTOR ROLE v3.5». ORGANIZATION ARTIFACT (OA) – used to specify a concrete organization artifact instance of a particular OAK. Example: «A01 - rental starter» is an instance of ELEMENTARY ACTOR ROLE v3.5. ORGANIZATION ARTIFACT RELATION (OAR) – used to specify a concrete organization artifact relation instance of a certain OAKRK. Example: «CAR PICK UP.executed by.RENTAL STARTER». ORGANIZATION ARTIFACT KIND ELEMENT KIND (OAKEK) – used to specify an element kind of an organization artifact kind. Example: Transaction Name is an element of the Organization Artifact Kind Transaction. ORGANIZATION ARTIFACT ELEMENT (OAE) – used to specify instances of an OAKEK. Example: the string "rental start" hat is the name given to a particular transaction that is an instance of the Organization Artifact Kind Transaction.

## 4 From DEMO to BPMN Workflow with Precise Semantics

The major goal of the DEMOBAKER project is to allow the conversion of UEAOM based DEMO models into compliant UEAOM based BPMN processes. Looking at DEMO models, we concluded that Action Rules would be the main source of information for this conversion process, as they specify, for all transactions, all agendum for each of the internal actors of the organization, that is all coordination facts that they have to respond to and then which conditions have to be verified, facts created, etc. After several experiments of converting DEMO models to BPMN models we found that we had to discard several ambiguous elements from BPMN – e.g., the message element that easily becomes redundant with tasks – and arrived at the following conversion rules – one of the main contributions of this paper – so that each DEMO concept has a 1 to 1 correspondence to a BPMN concept. We use the format: **DEMO concept < > BPMN concept**. Because DEMO has a strong semantics with a comprehensive meta-model, these proposed rules imply that, by using the few BPMN elements we select, we have a more precise semantics for BPMN compared to an unrestricted use of it or to using BPMN as a starting point to model enterprise processes.

**Transaction < > Pool** – each transaction is represented on BPMN as a Pool, and inside of this pool we can only have its related coordination and production acts/facts. Each transaction must start with a *start event* and finish with an *end event*. The start event must be triggered by another transaction. The end event can occur due to several reasons, for example, due to a revoke request act realized by a certain actor.

**Actors < > Lane** – Actors initiate and/or execute transactions, so each transaction has two lanes, one for the initiator and another for the executor and all events depicted inside a lane are of the responsibility of the respective actor.

**Flows and Conditions < > Tasks** – Flows, conditions and their evaluations (in action rules) are all represented on BPMN as tasks. Tasks have an input and an output flow. Each actor is responsible for the tasks depicted in their respective lane.

**Coordination-Facts/Production-Facts < > Signals** – Coordination and production facts correspond to signals, meaning that they are used as throw and catch signals. This is a key conversion rule and a very important contribution of our research. In this manner we can "isolate" the specification of each transaction and their respective rules in a pool and the occurrence of certain facts will possibly enact one or more transactions at the same time. This provides a high degree of flexibility, modularization and paralelism.

## 5 New DEMO Action Rule Syntax

As already mentioned, to present our proposal of a new DEMO Action Rules Syntax, we use the case EU-rent from [14]. In Figure 5 we find one of the action rules from this case. And this one is a perfect example on how Action Rules are "the neglected son" of DEMO. In real life, many different conditions and facts have to be verified before one can proceed to accept the drop-off of a car. In this action rule specification, the only verified fact is if the branch where the car is delivered is the same as the contracted one. But no action is specified for the case it is not. Also, in this rule we find a common problem in DEMO's Action Rule Meta-Model: what is the meaning of the construct *with*? We find it in many action rules and, apparently, with different functions: creating new facts, verifying new facts, etc. Indeed, in the most current public version of DEMO, it is assumed that "the syntax and the formal semantics of the action rules need to be elaborated yet" [22]. We agree that it is valuable to have models that abstract from infological and datalogical aspects as well as implementation issues, like DEMO's Construction Model, Process Model and State Model do. However, DEMO models can never be fully independent of implementation and resource constraints arising from the organization's reality. Rather, at most, they are implementation abstracted [23]. We claim that Action Rules are the perfect spot to make the bridge between the implementation world and the most implementation abstracted views of an organization – like that of transactions and actor roles of the Actor Transaction Diagram. We claim this because, while thinking on the flow and requirements for the action of actor roles, we inevitably need to think about necessary fact evaluations, information requests, data storing etc. So why not specify immediately such items while devising DEMO action rules? In this manner, the ontological model of an organization fully guides the specification of relevant infological, datalogical actions, as well as implementation requirements. We present, in Figure 6, the action rule that is the final result of the evolution of the simple action rule from Figure 5. This rule is already structured according to the new syntax we propose and is the result of several iterations of the next steps – that follow design science research method mentioned in Section 2: (1) devising new meta-model constructs for DEMO's Action Model and (2) evaluating their applicability and comprehensiveness by instantiating all the action rules of the EU-rent case. For step 1 we would create instances of UEAOM classes OAK, OAKRK and OAKEK, thus, specifying the meta-model. For step 2 we would create instances of UEAOM classes OA, OAR and OAE, that is, creating an Action Model following the specified meta-model. For space reasons we present only the action rule of Figure 6, which is already enough to explain and justify our proposal. This rule example shows many elements necessary for BPMN specification. The BPMN model resulting from our conversion rules – shown in Figure 7 – depicts the steps of the respective transaction and different kinds of infological and datalogical actions while still being totally abstracted from the implementation. The already mentioned design science research cycle of creation of, both the UEAOM instantiation for the meta-model specification and the instantiation for the full action rule specification of the EU-rent case, continued, until an instantiation gathered in a comprehensive way all the necessary information to have a real-life BPMN process fully specified, excluding implementation details. In our BPMN example we find all needed information, namely, all relevant ontological, infological and datalogical steps regarding the process of dropping the car at a branch.

| | |
|---|---|
| WHEN | car drop-of of [rental] is stated |
| with | the actual drop-off branch of [rental] is [branch] |
| then | car drop-of of [rental] must be accepted |

**Fig. 5.** Action Rule example

Figure 8 shows the UEAOM instantiation that consists in the new Action Meta-Model we propose and the explanation of each of its elements is now due. Due to the UEAOM following the AOM pattern and also the type square pattern, the explanation that follows could appear confusing, having too many instantiations. But the reader just needs to keep in mind that these patterns provide immense power of adaptability to systems in runtime due to the fact that instances of classes of an AOM may be themselves types or "classes" and that instances of AOM classes may, in turn, have instance kind relationships between them.

The first instance of the UEAOM class OAK that we need is, of course, the one specifying the *action rule* concept type itself. It has an associated OAKEK instance for its identification that we call *action rule id*. In our example, an instance of the UEAOM class OA would be action rule *AR05*. Being that the AR05 string or value is an instance of an OAE, itself instance of the OAKEK action rule id. This rule AR05 is, itself, an instance – at model level – of the just before mentioned OAK instance action rule – in turn, at meta-model level. To the reader that had no knowledge of the type square pattern. These several "double instantiations" described in the previous sentences constitute an example of 4 AOM instances/objects, forming such a type square, where, on one side we have a type and a property type (part of the type) and on the other side of the square we have an instance of the type containing a value, itself instance of the respective property.

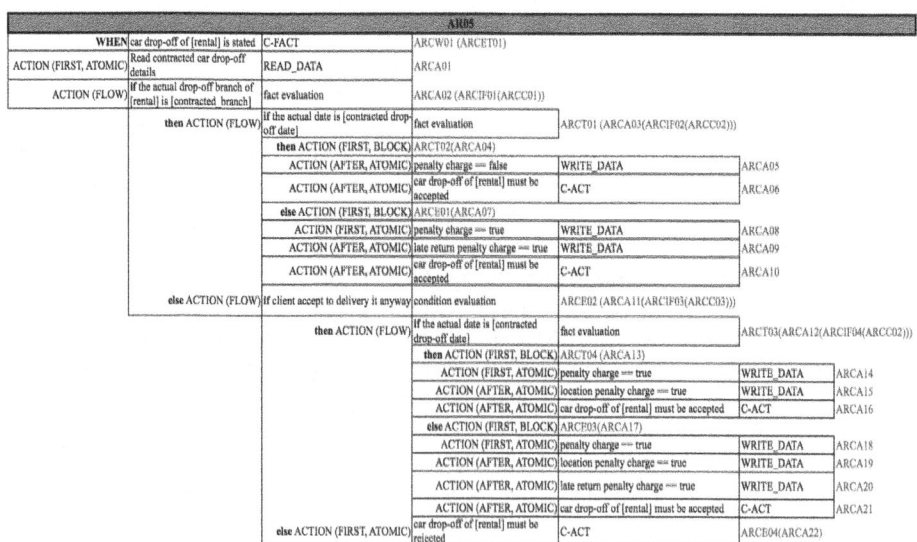

**Fig. 6.** Action Rule example following the new syntax

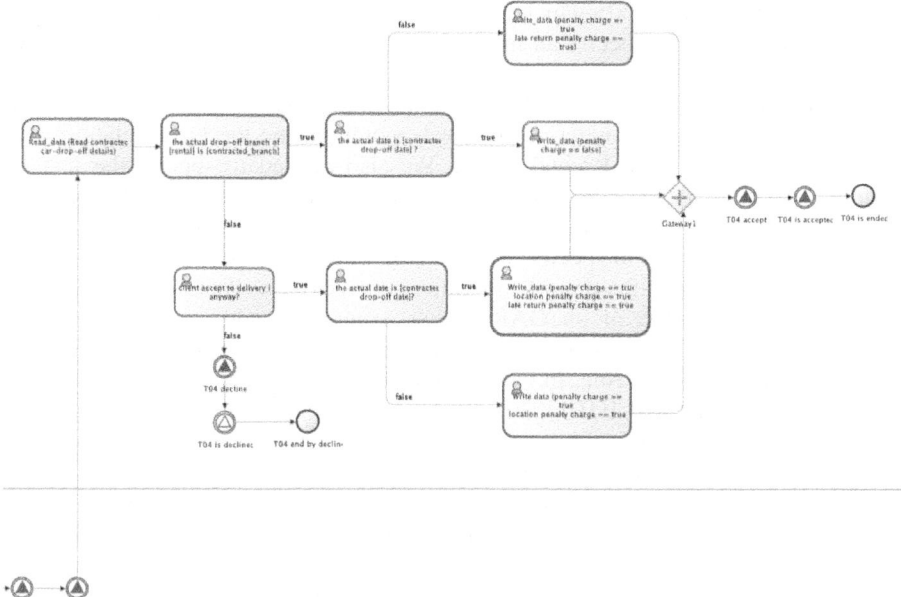

**Fig. 7.** BPMN diagram for Transaction T04 - car drop off

| | | |
|---|---|---|
| Organization Artifact Kind | **Action Rule** | |
| OAK Relation Kind | **executing actor** | |
| OAK Element Kind | action rule id | <string> |
| Organization Artifact Kind | **AR-Component-WHEN** | |
| OAK Relation Kind | **when component part of action rule** | |
| OAK Element Kind | arcw-id | <string> |
| Organization Artifact Kind | **AR-Component-ENACTING TRANSACTION** | |
| OAK Relation Kind | **enacting transaction component part of when component** | |
| OAK Relation Kind | **enacting transaction** | |
| OAK Element Kind | enacting transaction C-fact | <requested, promised, stated, accepted, revoke request requested, revoke resquet allowed, revoke request refused, revoke promise requested, revoke promise allowed, revoke promise refused, revoke statement requested, revoke statement allowed, revoke statement refused, revoke acceptance requested, revoke acceptance allowed, revoke acceptance refused, rejected, declined> |
| OAK Element Kind | arcet-id | <string> |
| Organization Artifact Kind | **AR-Component-CONDITION** | |
| OAK Relation Kind | **subcondition of condition** | |
| OAK Element Kind | condition type | <AND, OR, NOT, EXPRESSION> |
| OAK Element Kind | expression | <string> |
| OAK Element Kind | arcc-id | <string> |
| Organization Artifact Kind | **AR-Component-IF** | |
| OAK Relation Kind | **flow component of action** | |
| OAK Relation Kind | **evaluated condition** | |
| OAK Element Kind | arci-id | <string> |
| Organization Artifact Kind | **AR-Component-FOREACH** | |
| OAK Relation Kind | **flow component of action** | |
| OAK Relation Kind | **type of element** | |
| OAK Element Kind | arcf-id | <string> |
| Organization Artifact Kind | **AR-Component-WHILE** | |
| OAK Relation Kind | **flow component of action** | |
| OAK Relation Kind | **evaluated condition** | |
| OAK Element Kind | arcwhile-id | <string> |
| Organization Artifact Kind | **AR-Component-THEN** | |
| OAK Relation Kind | **then component part of if component** | |
| OAK Element Kind | ar-ct-id | <string> |
| Organization Artifact Kind | **AR-Component-ELSE** | |
| OAK Relation Kind | **else component part of if component** | |
| OAK Element Kind | arce-id | <string> |
| Organization Artifact Kind | **AR-Component-ACTION** | |
| OAK Relation Kind | **action component part of component** | |
| OAK Relation Kind | **previous action** | |
| OAK Relation Kind | **fact creation** | |
| OAK Element Kind | precedence type | <FIRST, AFTER> |
| OAK Element Kind | action type | <ATOMIC, BLOCK, FLOW> |
| OAK Element Kind | atomic action type | <WRITE_DATA, READ_DATA, TRANSMIT_DATA, SPECIFY_DATA, PRODUCE_DATA, PRODUCE_DOC, COPY_DOC, STORE_DOC, GET_DOC, READ_DOC, C-ACT, P-ACT> |
| OAK Element Kind | action description | <string> |
| OAK Element Kind | generic requirement | <string> |
| OAK Element Kind | implementation requirement | <string> |
| OAK Element Kind | arca-id | <string> |

**Fig. 8.** UEAOM based Action Rule Meta-Model specification

Next, we needed to specify which actor role has the responsibility of executing this rule. Thus we needed to specify an instance of class OAKRK, that we call *executing actor*, relating the action rule OAK with the actor role OAK. In our example, an instance of an OAR that would be an instance at model level of the OAKRK executing actor (in turn, at meta-model level) would be the the OAR relating AR05 with actor role A01, named rental starter. We also needed to specify the fact that triggers an action rule. For that we specified the OAK *AR-Component-When*, also needing an identifier, specified as the OAKEK *arcw-id*. In our example the instance is ARCW01. Next, we need to specify to which action rule this AR-Component-When is related to. Thus the specification of the following OAKRK: *when component part of action rule*. An instance of an OAR in our example would be the one relating ARCW01 to AR05. The OAK *AR-Component-Enacting-Transaction* serves to specify facts that trigger an action rule. In our example, an OA instance that is an instance of this OAK instance is: ARCET01. This complexity is needed as an action rule can be triggered by one or more c-facts of different transactions. The OAKRK *enacting transaction component part of when component* serves to relate the previous OAK with the *when component* OAK. In our example, an instance of an OAR instance of this OAKRK would be the OAR relating ARCET01 with ARCW01. The OAK Relation Kind *"enacting transaction"* specifies the relation between the OAK *AR-Component-Enacting-Transaction* and the OAK transaction. In our example, an instance of an OAR instance of this OAKRK would be the OAR relating ARCET01 with transaction T04 named car drop-off. The OAKEK *"enacting transaction C-Fact"* serves to specify which fact (out of the possible 21 facts of the universal transaction pattern) of the related transaction triggers the execution of this action rule. In our example we have the OAE *stated*. All the above instances of classes OAK, OAKRK and OAKEK are needed to precisely specify, at meta-model level, the structure of the initial part of an action rule. The above mentioned instances of classes OA, OAR and OAE specify at model level the following part of our example action rule: *"when car drop-off of [rental] is stated.*

Similar reasonings were followed for all other OAKs, OAKRKs and OAKEKs that we specify and are presented in Figure 8. We now only justify their specification. We need to specify the OAK *condition*. Conditions are evaluated by what we called *Flows* (If, While), explained later. A condition has to be of a certain (OAKEK) *condition type*, namely: *AND, OR, NOT*, or *EXPRESSION*. In the first three cases the condition is actually a composite condition where we have a boolean operator followed by a (OAKRK) sub-condition. In the last case the condition will actually be an atomic condition in the form of a boolean expression. In our example in Figure 6, we would have, as instance of class OA, the atomic condition: ARCC01 with expression *the actual drop-off branch of [rental] is [contracted_branch]*. In the case we have composite conditions (in fact a tree of conditions) each non-root condition needs to specify that they are *sub-condition of condition*. Imagining we would have a NOT condition with the previous example being the expression, we would have an OAR specifying that ARCC02 is a sub-condition of ARCC01, where the condition type OAE associated with the OA ARSC02 would be: NOT. Having the conditions specified we need to specify in which context they are evaluated. This happens in two kinds of *flow* component. Both the (OAK) *if* and the (OAK) *while* components evaluate a condition specified by the (OAKRK) *evaluated condition*. We specify to which action rule they belong thanks to OAKRK *flow component of action*. In our example we would have the OAR: ARCIF01 evaluating the condition ARCC01. The *foreach* flow has an action for each of its *type of element*. An if component is composed by its respective (OAK) *then* and (OAK) *else* components. In our example: the OAK ARCT01 being the *then component part of if component* ARCIF01 and the OAK ARCE01 as the *else component part of if component* ARCIF01. Finally we needed to specify actions of the action rule itself, thus the OAK AR-Component-Action. Actions always belong to some component, relation specified by OAKRK *action component part of component*. In our example: There is an OAR specifying the OA ARCA01 as being the *action component part of component* OA ARCW01 (the when component of action rule AR05). An action component can be followed by another action and such precedences are specified by the OAKRK *previous action*. In this manner we can "program" actions sequentially. For instance and based on our example: The OA ARCA05 that has as *previous action* the OA ARCA04. When an actor will perform an action, some fact can be created and that relation is specified by the OAKRK *fact creation*. Since actions can proceed other actions or may be the first action for a certain component, we had the need to be able to specify this differentiation. Thus we specified the OAKEK *precedence type* where the allowed values for the OAE are strings *FIRST* or *AFTER*. To specify if the action component in question is an atomic action or a block of actions or a flow we need OAKEK *action type*. In our example the OA ARCA04 is the FIRST action of the then component and is of type BLOCK, that is, it just specifies that we will have a block of actions executed. The OA ARCA05 is the FIRST ATOMIC action of the action block ARCA04. An atomic action can be of many types and this is specified by OAKEK *atomic action type*. As we can see in Figure 8 there are datalogical, infological and ontological acts as options. In our example: the OA ARCA05 has its OAE with value *WRITE_DATA*. OAKEK *action description* serves to specify/describe in a formal/informal way the atomic action in question. In our example we have the OAE *penalty charge == false*. An action can also have specific requirements still abstracted from implementation or implementation-dependent. Thus the OAKEKs *generic requirement* and *implementation requirement*. As an example of a generic requirement we could specify that a certain action of type PRODUCE_DATA "is mandatory" which means that the actor has to really get or produce such data from somewhere. An example of an implementation requirement would be "obtain record from the government driving license system's web service"

We could produce an OFD (the DEMO option for specifying meta-models) for another alternative view (other than the UEAOM instantiation) of our newly proposed meta-model for DEMO Action Rules. Due to lack of space and taking in account that we are kind of specifying what we could call the *organization programming language* we choose to present, in Figure 9, our newly proposed syntax using the Backus-Naur Form (BNF) notation, which summarizes our main contribution of this paper.

| | | |
|---|---|---|
| <When> | ::= | <Transaction kind name> "of" <Object Identifier> "is" <C-Fact> | <P-Fact> <Action> | <Flow> |
| <Transaction kind name> | ::= | <String> |
| <Object Identifier> | ::= | "[" <String> "]" |
| <P-Fact> | ::= | "produced" |
| <C-Fact> | ::= | "requested" | "promised" | "stated" | "accepted" | "revoke request requested" | "revoke resquet allowed" | "revoke request refused" | "revoke promise requested" | "revoke promise allowed" | "revoke promise refused" | "revoke statement requested" | "revoke statement allowed" | "revoke statement refused" | "revoke acceptance requested" | "revoke acceptance allowed" | "revoke acceptance refused" | "rejected" | "declined" |
| <C-Act> | ::= | <Transaction kind name> "of" <Object Identifier> "must be" <C-Fact> |
| <P-Act> | ::= | <Transaction kind name> "of" <Object Identifier> "must be" <P-Fact> |
| <Flow> | ::= | <Foreach> | <If> | <While> |
| <Action> | ::= | <Atomic action> | <Block> | <Flow> |
| <Atomic action> | ::= | <Action type> <Action description> | <Generic requirement> | <Implementation requirement> |
| <Action description> | ::= | <String> |
| <Generic requirement> | ::= | <String> |
| <Implementation requirement> | ::= | <String> |
| <Action type> | ::= | "Write_Data" | "Read_Data" | "Transmit_Data" | "Specify_Data" | "Produce_Data" | "Produce_Doc" | "Copy_Doc" | "Store_Doc" | "Get_Doc" | "Read_Doc" | <C-Act> | <P-Act> |
| <Block> | ::= | <Action> { <Action> } |
| <Condition> | ::= | <Condition type> { <Condition> } | <Expression> |
| <Condition type> | ::= | "And" | "Or" | "Not" | <Expression> |
| <Expression> | ::= | <String> |
| <If> | ::= | "If" <Condition> "then" <Action> | <Flow> "else" <Action> | <Flow> |
| <While> | ::= | "While" <Condition> <Action> | <Flow> |
| <Foreach> | ::= | "Foreach" <type of element> <Action> | <Flow> |

**Fig. 9.** New Action Rule Syntax

## 6 Discussion, Conclusions and Future Work

When one really starts to specify details, still at infological and datalogical levels, things can get really complex. By looking at the full model of the EU-Rent case [14], one can see that it's only on the rule that handles the promise of the rental end transaction, that the payment transaction is requested (possibly with incurring fines). Imagine that the renter would drop off the car by mistake in a branch different from the contracted one. After dropping off the car at the garage he or she would get quite a surprise, at the branch desk, when having to pay the fine. It makes much more sense that the action rule that handles the state c-fact of the car drop-off transaction evaluates if the drop-off branch is correct and already informs the renter of the fine he would have to pay if he wishes to proceed. This gives him or her a chance of not proceeding and maybe leave this branch and deliver the car on the correct branch. Just this example shows that, while modeling DEMO transactions, one should already have in mind implementation issues that will affect transaction design. Depending if (1) there is a garage attendant and then the rental desk handles the payment or if (2) there is just the garage attendant himself which takes care of both the car drop-off and penalty payments; this will have a profound impact on the design of the action rules and maybe even on transaction design itself. In this later case one could "fuse" the current car drop-off and rental end transactions. In the more complete action rule we present on this paper, we see that an apparently simple rule was in fact "forgetting" lots of complexity in terms of conditions to be verified and the creation of many original facts (e.g., flags regarding the fines) that are themselves, very relevant actions to be correctly and comprehensively implemented in a BPMN flow in the correct spot of the flow, like we depict in Figure 7. We started off with the purpose of automatically generating runnable BPMN models from DEMO models. With this end in mind, we had first to solve the problem of lack of semantics in BPMN. Hence, we selected a few BPMN concepts (out of the many ambiguous ones available) and assigned to them clear and precise semantics thanks to the the set of 1 to 1 conversion rules from DEMO to BPMN that we propose. But in the middle of the process we found yet another problem: the main information source in DEMO to generate BPMN flows – the Action Rules Specification of the Action Model – was not precise enough nor had clear semantics itself. We then took the endeavor of, following Design Science Research tenets, solving that problem by applying the Universal Enterprise Adaptive Object Model to specify a more complete and comprehensive Meta-Model, i.e., abstract syntax, for the Action Rules. After several instantiations of our ideas and evaluation of their applicability by the instantiation of all EU-rent case action rules with each new version of our Meta-Model, we kept on improving it to the stage as presented in this paper. Some very relevant contributions of the new syntax are that we are kind of finding the primitives for what we call *organization programming language*, still at an implementation abstracted level, in a way that we can produce much more comprehensive BPMN models, more close to being ready to be runnable, compared to other existing approaches. The current version of the syntax has several aspects to improve still. Namely when we have an atomic action that is a C-ACT, we must have an additional OAKEK to specify to which transaction this C-ACT corresponds, so that the Workflow engine can throw the right signal to activate the right BPMN pool. In fact one of the next lines of future work, besides polishing up missing details in our new syntax, is the creation of a parser that takes, as an input, action rules following our proposed syntax and outputs a runnable BPMN workflow that can be automatically imported to a well-known Open Source Workflow System where we can easily implement (or connect to) the needed database tables, queries and web forms for data production/input, following the requirements formally or informally specified in each atomic action of the action rules.

We also have one small contribution to the formal specification of the Universal Transaction Pattern. All coordination acts need to have different names. For example, in the current pattern, revoking a request is different from revoking a promise. So the allow and refuse acts – which in the current pattern have the same name for the 4 stanard c-acts – should also be differentiated according to if they correspond to allowing the revoke of a request or of a promise, for example. All the new names of c-acts we propose can be found on our new syntax and the official transaction pattern should have the names of these same acts changed accordingly.

# References

1. Dalal, S., Chhillar, R.S.: Case Studies of Most Common and Severe Types of Software System Failure. International Journal of Advanced Research in Computer Science and Software Engineering 2 (2012)
2. Shull, F., Basili, V., Boehm, B., Brown, A.W., Costa, P., Lindvall, M., Port, D., Rus, I., Tesoriero, R., Zelkowitz, M.: What We Have Learned About Fighting Defects. In: Proceedings of 8th International Software Metrics Symposium, pp. 249–258 (2002)

3. Zeller, A., Hildebrandt, R.: Simplifying and Isolating Failure–Inducing Input. IEEE Transactions on Software Engineering (2002)
4. Dalal, S., Chhillar, R.S.: Role of Fault Reporting in Existing Software Industry. CiiT International Journal of Software Engineering and Technology (July 2012)
5. Dietz, J.L.G.: Enterprise Ontology: Theory and Methodology. Springer, Heidelberg (2006)
6. Dumay, M., Dietz, J.L.G., Mulder, H.: Evaluation of DEMO and the Language/Action Perspective after 10 years of experience. In: Proceedings of LAP 2005 (2005)
7. Aveiro, D., Silva, A.R., Tribolet, J.: Towards a G.O.D. Organization for Organizational Self-Awareness. In: Albani, A., Dietz, J.L.G. (eds.) CIAO! 2010. LNBIP, vol. 49, pp. 16–30. Springer, Heidelberg (2010)
8. Aveiro, D., Rito Silva, A.: Extending the Design and Engineering Methodology for Organizations with the Generation Operationalization and Discontinuation Organization. In: Winter, R., Zhao, J.L., Aier, S. (eds.) DESRIST 2010. LNCS, vol. 6105, pp. 226–241. Springer, Heidelberg (2010)
9. The Open Group: ArchiMate® 2.1 Specification, http://pubs.opengroup.org/architecture/archimate2-doc/
10. Object Management Group: BPMN 2.0, http://www.omg.org/spec/BPMN/2.0/
11. Dijkman, R.M., Dumas, M., Ouyang, C.: Semantics and analysis of business process models in BPMN. Information and Software Technology 50, 1281–1294 (2008)
12. Ettema, R., Dietz, J.L.G.: ArchiMate and DEMO – Mates to Date? In: Albani, A., Barjis, J., Dietz, J.L.G. (eds.) CIAO! 2009. LNBIP, vol. 34, pp. 172–186. Springer, Heidelberg (2009)
13. Aveiro, D., Pinto, D.: Universal Enterprise Adaptive Object Model. Presented at the 5th International Conference on Knowledge Engineering and Ontology Development (KEOD), Vilamoura, Portugal (September 2013)
14. Jan, L.G.: Dietz: DEMO 3 - Way of Working (2009)
15. Hevner, A.R., March, S.T., Park, J., Ram, S.: Design science in information systems research. Management Information Systems Quarterly 28, 75–106 (2004)
16. Hevner, A.: A Three Cycle View of Design Science Research. Scandinavian Journal of Information Systems 19 (2007)
17. Dietz, J.L.G.: Is it PHI TAO PSI or Bullshit? Presented at the Methodologies for Enterprise Engineering Symposium, Delft (2009)
18. Dietz, J.L.G.: On the Nature of Business Rules. In: Dietz, J.L.G., Albani, A., Barjis, J. (eds.) CIAO! 2008 and EOMAS 2008. LNBIP, vol. 10, pp. 1–15. Springer, Heidelberg (2008)
19. Dietz, J.L.G., Albani, A.: Basic notions regarding business processes and supporting information systems. Requirements Eng. 10, 175–183 (2005)
20. BPM Tutorial - TechTarget, http://searchsoa.techtarget.com/tutorial/BPM-Tutorial
21. Yoder, J.W., Balaguer, F., Johnson, R.: Architecture and design of adaptive object-models. SIGPLAN Not. 36, 50–60 (2001)
22. Dietz, J.L.G.: DEMO-3 Models and Representations (2009), http://www.demo.nl
23. Pombinho, J., Aveiro, D., Tribolet, J.: The Role of Value-Oriented IT Demand Management on Business/IT Alignment: The Case of ZON Multimedia. In: Harmsen, F., Proper, H.A. (eds.) PRET 2013. LNBIP, vol. 151, pp. 46–60. Springer, Heidelberg (2013)

# Detailed Analysis of REA Ontology

Frantisek Hunka and Jaroslav Zacek

University of Ostrava, Department of Computer Science
Dvorakova 7, 701 03 Ostrava 1, Czech Republic
{frantisek.hunka,jaroslav.zacek}@osu.cz

**Abstract.** The paper addresses REA (Resource-Event-Agent) domain specific ontology that is primarily focused on value modeling in business processes. REA ontology which historically originates from accounting information systems, gradually developed to cover all areas where value modeling can be utilized. After a short introduction, a core REA pattern is introduced and analyzed from the view of its basic entities and the relationships between them. Next, additional crucial concepts of REA ontology and their relationships are gradually elucidated and analyzed from the view of DEMO (Design & Engineering Methodology for Organizations). The paper also describes the current definition of economic transaction that is used as a basis for REA state machine. The discussion and conclusion sections summarize and assess the pros and cons of REA ontology and propose a way forward for further research.

**Keywords:** business process modeling, value modeling ontology, REA core pattern, REA value model.

## 1 Introduction

The REA model originated from the accounting domain and matured to a conceptual framework and ontology for Enterprise Information Architectures [2, 3]. This data modeling perspective allows the inclusion of structural and behavioral aspects of business objects within the data model. Business objects are then represented in information systems according to specified models.

The REA model provides concepts to store past and future data consistently [12]. It also provides concepts to explicitly define business processes in the same framework as business event data [12]. The REA model records information based on the coherence between the data of one or more business events. The REA process is defined by related REA events and has at least two composite economic events: a decrement event that consumes or uses the incoming resource(s) and an increment event that acquires the outgoing resource(s). REA process is called REA value model and represents the notion of business process. REA value models can be bound into an REA value chain which usually creates a closed cycle. While REA process is defined by related REA events, REA value chain is specified by related REA value models.

REA ontology benefits from the presence of a semantic and application independent data model, an object oriented perspective, and abstraction from technical and implementation details. These features enable the possibility to calculate the value of the enterprise's resources on demand as opposed to calculation at pre-determined intervals.

An REA value model is designed to input and output a resource entity. That is in accordance with the REA value chain concept that is defined as a set of business processes through which resources flow [5]. The REA value modeling paradigm is principally based on the observation of resource value entities. This is why both concepts, the REA value model and the REA value chain, represent different levels of abstraction, and that only a resource's flow is possible to flow between two business processes. REA ontology is foremost connected with REA value model and its models belong to a category of data models.

The motivation of the analysis may be twofold. Firstly, the analysis of REA framework is the initial step in revealing deficiencies and incompleteness (even misconception) in the REA value modeling approach which may be a guideline for gradual improvement of the REA framework. Secondly, the analysis may leads to findings concerning DEMO possibilities to be closer utilized in compliance with other domain specific anthologies.

The remainder of the paper is organized as follows: REA core pattern and REA value chain are described in section 2. The commitment entity that deals with future activities is characterized in section 3. The type level of the REA value model is depicted and assessed in section 4. The business transaction that is utilized as a basis of the REA state machine is clarified in section 5. REA approach example is depicted in section 6. The resultant findings are discussed in section 7 and the final section contains conclusions and a summary of future work to be performed.

## 2   REA Core Pattern

REA ontology is based on the REA core pattern that expresses the basic principle. This core pattern comes from double booking accounting systems and relates to exchange phase of economic resources. The first finding that was identified by [11] was a relationship between a pair of economic events as a part of an exchange. These different events were involved in various transactions, they had something in common; there was always a decrement economic event (one in which something is provided) and an increment event (one in which something is received). These events could be paired. The other feature of these events was that exchange was not always immediate and there were no rule as to whether the increment or the decrement event occurred first. It happened that there was a significant time lag between the events, for which double booking entries created an account to allow the entries to balance. These timing differences were called *claims* by [11] and can be recognized as accounts receivable, deferred revenue, prepaid expenses, accounts payable, and wages payable. Exchanged items always have an economic value and could be thought of as resources. The other finding made by [11] was that each event making up the exchange also involved human beings whom he called economic agents. The REA core pattern is illustrated as a Pizzeria shop in Fig. 1.

Economic events in the exchange processes represent the permanent or temporary transfer of rights to economic resource from one economic agent to another. The transfer of rights represents the increment or decrement of the value of the resources. In short, the purpose of an economic event in the REA exchange process is to transfer some of the rights associated with the resource from one economic agent to another. The economic events in REA application models usually encapsulate properties for *date* and *time* and *location* in space. Recognition that an enterprise's economic activities follow REA core pattern in which causally related *give* and *take* events are associated with resources and agents was made by [11].

Economic events and the duality relationship by which events are related play a crucial role in this pattern. All other entities that participate in this pattern are related 'through' economic events. Both of agents are related with increment and decrement events because they lose rights to given resources and gain rights to other resources. On the other side, resources are related to corresponding economic events. The aim of the *claim* entity is to balance any inequalities that occur.

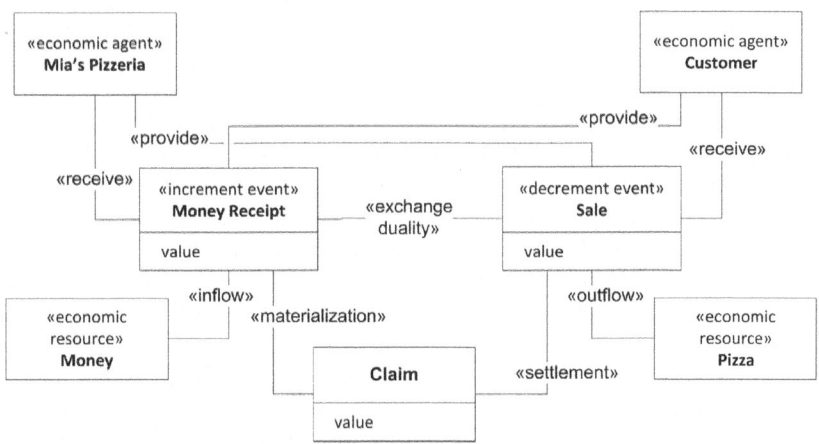

**Fig. 1.** REA core pattern

The relation between increment and decrement events is exchange duality. The purpose of exchange duality is to keep track of which resources were exchanged for which ones. In the REA value model of an exchange process, every increment economic event must be related by an exchange duality to a decrement economic event, and vice versa.

Apart from exchange process, REA ontology also distinguishes conversion process of which economic resources can be used, consumed or produced in order to create new economic resources or to change features of existing resources by using or consuming resources of the same or of another kind. Economic events in the conversion processes can change the value of the features, as well as add and remove features to and from resources. Conversion duality is the relationships among corresponding economic events within the conversion process.

However, as the aim of the paper is to analyze the principal issues of the REA ontology, we will no longer devote ourselves to the conversion process as the difference between an exchange and conversion processes is not principal.

One of the important characteristics of economic events in REA ontology is that economic events can register only those events that have already occurred or are occurring in the present. Economic events can certainly be planned or expected to occur in the future; the REA concept of commitment describes the events that have not yet occurred. Commitments register economic events that will occur in future, as will be seen in section 3.

A business process in REA ontology is characterized as paired transactions such as *sale-money receipt*. These parts constitute a business process. In general, REA ontology provides a 'dependent' view of this process. It means that the process is viewed from the side of one of these economic agents. In this case, one of these 'transactions' can be considered to be the main transaction, while the other can be considered to be enclosed within the main 'transaction'.

The claim entity that balances the discrepancies among economic events has no further specification apart from the relationships of materialization and settlement by which it is related to economic events.

In a closer look at an REA economic event with respect to possible similarities to the DEMO basic transaction pattern, we can see that the REA core pattern is comprised of two human beings called economic agents between them the exchange (conversion) is held. The core pattern is represented by a pair of 'transactions' related by the duality relationship between corresponding event entities. The event entity by its relationships of provide and receive has similarities to the *result phase* of the DEMO basic transaction pattern. The provide relationship has some analogy with the state process step and the receive relationship has some analogy with the accept process step. In addition, the event entity includes the properties of date, time and location. From this point of view, the economic event also records the data of the execution phase of the DEMO basic transaction pattern, so we can deduce that the event entity also tacitly meets the prerequisites of the *execution phase* of the DEMO basic transaction pattern.

The REA core pattern fully reflects the idea of an accounting system. It only registers past and present events in the view of resource value changes which happened or has happened. Unfortunately, the future events are out of the scope of the REA core pattern.

## 2.1 Value Chain Concept

As we stated before, the REA value model and REA value chain constitute the main building blocks of the REA ontology. The REA value chain put together REA value models to create higher abstractions for enterprise modeling. The meaning of the REA value chain can be exemplified in e.g. Mia's Pizzeria, where several exchange and conversion processes, the Sales, Purchase, Acquisition and Pizza Production can be identified. The REA value chain is created by flows of economic resources between individual processes. At the output of each process there is an economic resource that is an input of another process, see Fig. 2.

The pizza production process produces pizza, which is exchanged in the sales process for money. Mia's Pizzeria use money to purchase raw material and labor in the purchase and labor acquisition processes. The raw material and labor are consumed to produce pizza in the pizza production process. The REA value chain provides only sequence ordering of REA value models which is similar to the other traditional business process methodologies.

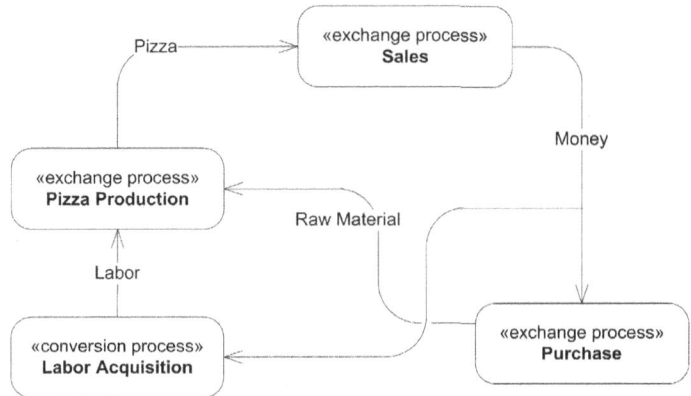

**Fig. 2.** Value chain of Mia's Pizzeria. Adapted from [7]

Although the REA value chain enables to model the whole application, there are still issues not satisfactory resolved yet. One of them is that the flow between REA value models is restricted to the resources and instances of resource's subclasses only. This limitation means that other REA value model entities such as a contract or a schedule that don't have inheritance relation to the resource entity, can't be easily transferred between REA value models within the REA value chain. This results in impossibility to create such entities in the REA value model and their utilization in other REA value models.

## 3  Future Activities and Commitment Entity

Commitment entity addresses the issue of modeling promises of future economic events and the issue of reservation of resources. The reason for this solution is that economic events specify according REA ontology only actual increment or decrement of resources, not the future increment or decrement of resources.

Commitment entities and their relationships with other entities are shown in Fig. 3. In this case, Fig. 3 is an extension of Fig. 1. The commitment entity copies to a considerable extent the structure of the event entity, by which we mean the existence of an increment and decrement commitment and exchange reciprocity relationship.

The exchange reciprocity relationship between the increment and decrement commitments identifies which resources are *promised* to be exchange for which others. The reciprocity relationship is a relation many-to-many (1..*, 1..*).

Each commitment is related to an economic resource by a reservation relationship which specifies what resources will be needed or expected by future economic events. The reservation relationship between the resource and commitment represents the features of the resource and rights associated with the resource that will be changed or transferred by a future economic event.

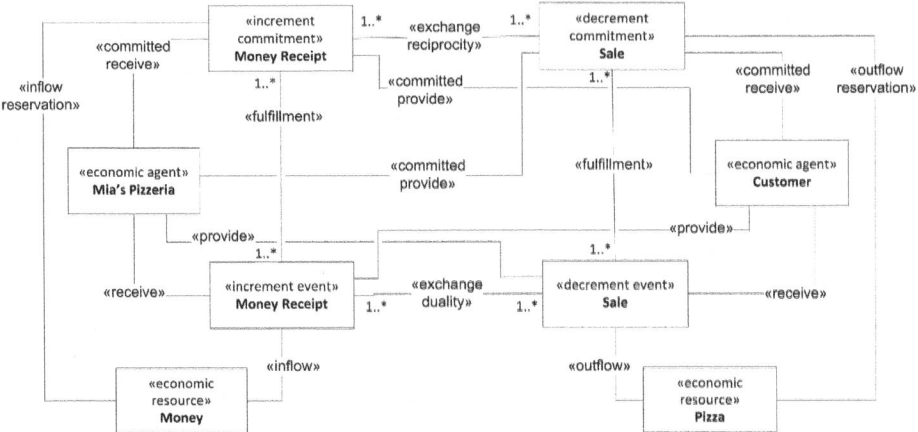

**Fig. 3.** REA value model with commitment and claim entities. Adapted from [7]

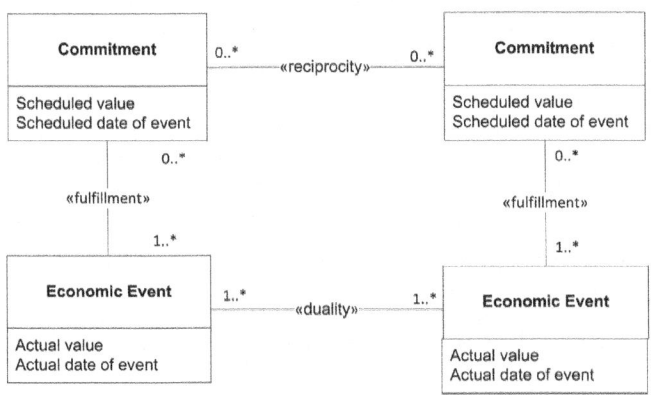

**Fig. 4.** Commitments and economic events. Adapted from [1, 7]

As a matter of fact, each commitment must be related by committed provide and committed receive relationships to economic agents.

The other crucial relationship is the fulfillment relationship which relates a corresponding commitment entity to an event entity. Fig. 4 shows the detail of this relationship. The purpose of this relationship is to validate whether the economic events fulfill corresponding commitments. This relationship is a many-to-many relationship. This means that one commitment can be fulfilled by several economic events, just as one shipment commitment can be fulfilled by partial shipments. Conversely, one economic event can fulfill several commitments, just as several installments can be paid at once. This functionality truly depicts established practice.

The principal feature of REA ontology is that it explicitly distinguishes between past and current events and events performed in the future for which it introduces the commitment entity. This explicit distinction between these two types of events has its origin in the gradual development of the REA framework. The other specific feature of the REA framework is a pair of 'transactions' which results in mirroring the corresponding commitment and event entities, see Fig. 4. The relationship of *committed provide* and *committed receive* mean that some agreement about the future exchange have to be achieved. From this point of view can be judged that the commitment entity in its essence complies with the *proposition phase* of the DEMO basic transaction pattern.

In general however, commitment and event entities lack states and state transitions. There were some attempts [4, 8] to introduce states and state transitions into REA ontology that utilized the business transaction standard [10] but have not been completely and clearly finished yet. The business transaction standard will be analyzed in section 5.

## 4 Type Level

REA value model explicitly utilizes types, which are used to define abstractions of economic phenomena such as *resource type*, *agent type*, *event type*, *commitment type* and additionally *contract type*, by [3]. E.g. *employee type* refers to a category of people who have the employment relationship with one or more organization. In an object oriented perspective, this construct was described as power types, see [8]. In general, there are two reasons for explicitly introducing this construct into REA ontology. Firstly, type definition, specifications and guidelines can be applied to each instance that conforms to the type. Secondly, the type definition enables the declaration of a type entity instance that holds all properties which create a "form" for all object instances that conform to the type instance. An example where the type construct is utilized is a resource type entity. All resource entities that conform to resource type entity share the properties of the resource type. In some activities where one is not able to specify actual resource instance, it is possible to use the resource type instance that fulfills all requirements. This type instance can be used e.g. in planning where a *reserve* relationship is utilized between commitment and resource type entities to keep them for future use.

As was mentioned before, the first reason for introducing types into the REA value model is the application of specifications and guidelines. REA ontology distinguishes among the following three types of policy definitions: *knowledge-intensive description*, *validation rules* and *target descriptions* see [3]. A *knowledge-intensive description* defines characteristics of a concept that apply to a group of objects.

*Validation rules* represent permissible values, and a common application of validation rules in enterprise systems are preventive controls. *Target descriptions* provide benchmarks regarding economic phenomena, and they can take at least two different forms: standards and budgets. Standards often refer to engineering information; budgets provide quantified performance measures most often related to specific time period. Applying types in REA value model is motivated by the needs to apply business rules in a broader context. There is no explicit model such as the DEMO's action model dealing with the rules and actions in REA ontology.

The second reason concerns the type definition that is applied especially to resources, events and agents entities. Resources are often categorized based on technical specifications, such as an octane rating for gas. Events are often categorized based on their method of execution, such as the mode of sales: distributor sales, direct sales, internet sales. Another example is the method of payment: cash, check, credit card. Agents can be categorized based on skills or roles. In practice however, the event type is utilized only rarely. The reason for this is brought about by the fact that event entity contains only actual information and all planned information are placed in the commitment entity. In this case, a commitment entity can be considered to be a "type entity" of the event entity. In terms of agent entity, the notion of *actor role* introduced in DEMO is in conformity with the *agent type* concept as opposed to the original *agent type* notion in REA ontology. The only meaningful type that is worthy of consideration is a resource type, rationale of which was previously described.

Commitments stated up till now represent the optimistic path of exchange. However, sometimes goods are not delivered as expected and payments arrive late. Partners usually agree upon what should happen if the initial commitments are unfilled. This is resolved by a *contract entity* that contains increment and decrement commitments that promise an exchange of economic resources between agents, and *terms* that specify additional commitments in the case of pessimistic path of exchange. Fig. 5 illustrates the *contract entity* and its relationships to other entities.

The figure contains only relevant entities in order to maintain clarity. *Terms* are potential commitments that are instantiated if certain conditions are met. These conditions can be various, such as a commitment not being fulfilled or a resource not being at a certain location. Economic agents usually agree upon penalties if the commitments are not fulfilled. If the commitments are unfulfilled, the *contract* will instantiate a new commitment to settle the discrepancies (pay a penalty). The *terms* and *commitments* are the *clauses* of the contract. Every *contract* must be related to two or more economic agents by a party relationship. These agents do not necessary have to be the provider and recipient of economic resources. The economic agents that comprise contractual parties can be different from the economic agents participating in the economic events which fulfill these commitments.

As REA ontology doesn't contain a similar model to the DEMO Action Model, the term entities are expressed in the form of on demand 'created' commitment entities that express 'unsuccessful path' of the process.

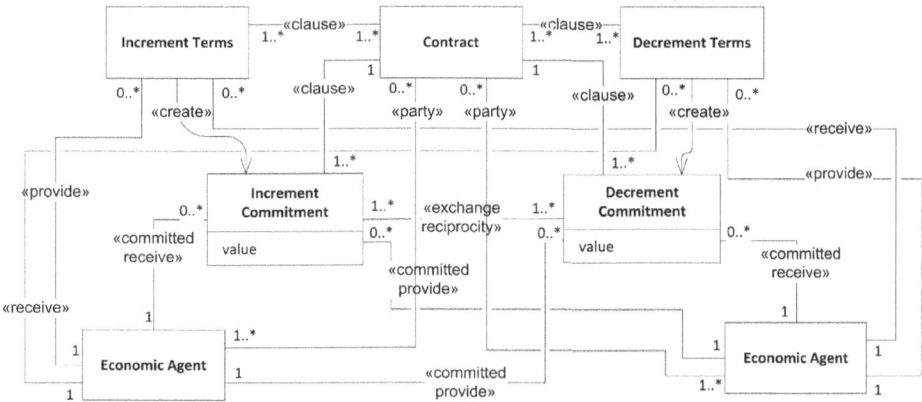

**Fig. 5.** Contract, commitments and terms. Adapted from [7]

As was mentioned earlier, declaring a contract entity without any relation to the resource entity brings about difficulty in creating the contract entity within the scope of the REA value model and consequently difficulty in the contract entity flow in the REA value chain (the same holds for the schedule entity too).

## 5 Business Transaction

REA ontology links its own states with the phases of the business transaction (sometimes labeled as business process). The business transaction is defined by ISO International Standard [9] and the definition is as follows:

*"Business transaction is a predefined set of activities and/or processes of Persons which is initiated by a Person to accomplish an explicitly shared business goal and terminated upon recognition of one of the agreed conclusions by all the involved Persons although some of the recognition may be implicit"* by [9].

Business transaction is taken as a reciprocal transaction like a transaction in the REA core pattern. According to the same standard, business transaction defines five individual phases:

- *The planning phase* serves for the buyer and seller (actor roles) to decide what action to take for acquiring or selling a good, service and so on.
- *The identification phase* represents interchange data and establishing one-to-one linkage.
- *The negotiation phase* pertains to the exchange of information following the identification phase. This phase leads also to the identification of each other at a level of certainly and at achieving an explicit mutual understanding. The process of negotiation is directed at achieving an explicit, mutually understood, and agreed upon goal of a business collaboration and associated terms and conditions.
- *The actualization phase* pertains to all activities for the execution of the results of the negotiation for an actual business transaction.
- *The post-actualization phase* includes all activities that occur between the buyer and the seller after the exchange.

Applying the Performa-Informa-Forma analysis on the individual phases, the following results are obtained. The *planning phase* contains foremost informational production. The *identification phase* from its essence represents foremost documental production and formative level of coordination. The *negotiation phase* is composed of informational production and the informative level of coordination and performative level of coordination, the process steps of *request* and *promise*. One can thus judge from the assertion that the parties agreed upon the goal of business collaboration. The *actualization phase* represents original production. The *post-actualization* phase contains the performative level of coordination, the process steps of state and accept.

Despite the fact that it is only a shallow analysis it reveals that the phases of the business transaction cover all basic steps of the DEMO transaction pattern.

All relevant entities and relationships between them are illustrated in Fig. 6 by [1]. Economic agreement represents either a *contract* or a *schedule* according to whether it is an exchange or a conversion process. The business transaction itself is governed by the economic agreement and serves as an aggregate of both a bundle of economic events bound by a duality relationship and a bundle of business transaction phases.

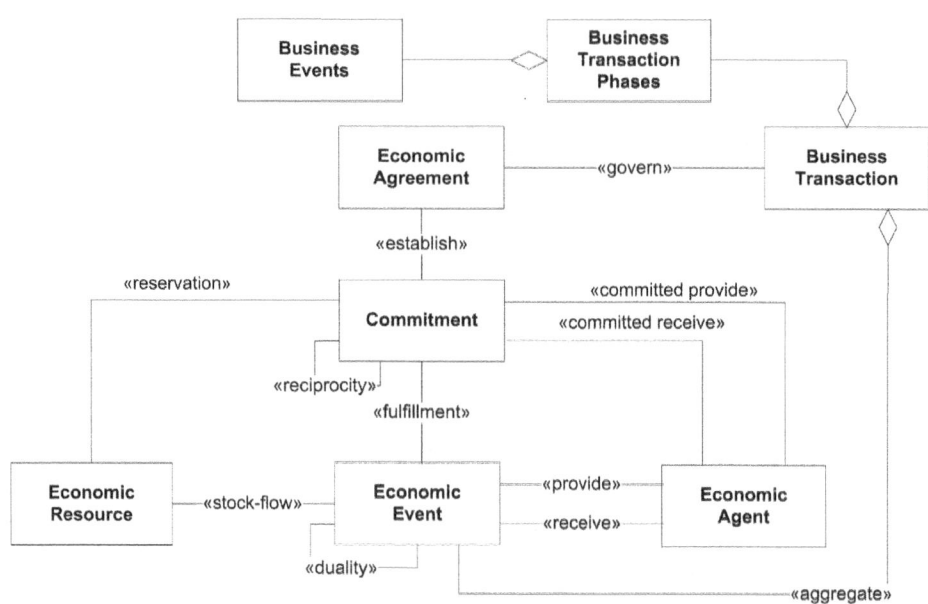

**Fig. 6.** Fundamental REA concepts including business transaction. Adapted from [1]

REA value model doesn't have implicitly declared states and state transitions so the business transaction phases are utilized explicitly to create REA state machine. However, states and state transitions should be identified and declared inside the REA value model mainly within the commitment and event entities. Better conceptual specification of the commitment entity and the event entity could lead to more rigorous definition of the business transaction and precise determination of the states and state transition within the scope of the REA value model.

## 6 REA Approach Example

This section illustrates practical utilization of the REA value framework. REA approach is exemplified by a REA value chain and a REA value model of an enterprise that we modeled. Fig. 7 illustrates the REA value model of all core processes of the enterprise. Resource flows link all corresponding REA value models. The figure contains both exchange and conversion processes. There are two planning processes. The first one is for a schedule creation (Planning Process Schedule), the second one is utilized for purchase order (Planning Process Contract) creation. The *planning process schedule* is a conversion process whose aim is to produce Production Schedule that is utilized both in the *production process* and in the *planning process contract*. The output entity of these processes represents schedule (contract) entity and a resource entity which stands for the value of the schedule (contract). The financing process supplies all dependent processes with money.

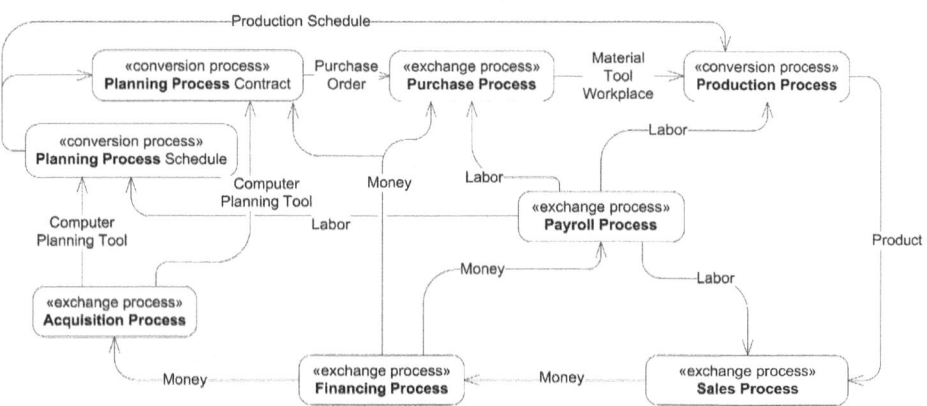

**Fig. 7.** REA value chain – practical example

Each of these processes can be illustrated in a more detail way using REA value model. More interesting processes are planning processes as they not only create a standard resource entity but also an information or knowledge entity in the form of a schedule or contract. The planning process of a schedule entity creation is shown in Fig. 8. The production order entity contains decrement commitments of *use of standards and directives*, *use of bill of material*, *labor consumption*, *use of computer* and *use of planning tools*. The last three decrement commitments are drawn in one box to make the figure more transparent. There is only one increment commitment of *create schedule* that is contained in the production order. As can be seen from the figure, increment and decrement commitments are related through reciprocity relationship. Each commitment is related to a resource type entity by a reservation relationship. The meaning of this relationship is to reserve corresponding kind of a resource type entity.

Each commitment is related to its corresponding event entity by the fulfillment relationship. Decrement events are related to the increment event by the conversion

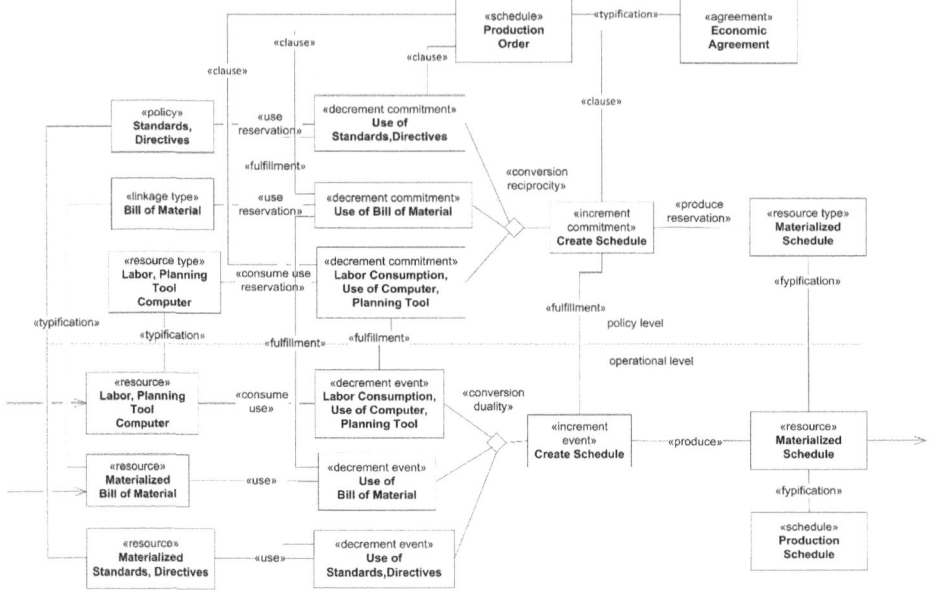

**Fig. 8.** Planning process schedule

duality relationship. Resource entities at the left hand side are either use or consumed during the conversion process. Typification relationship relates corresponding resource types to resource entity. The exact meaning of this relationship is explained in the chapter 4 – Type Level.

At the bottom of the right hand side of the figure, the materialized schedule entity and production schedule entity are produced. The materialized schedule entity represents the value of the production schedule entity.

The practical example of the REA value approach not only shows possibilities of the REA framework for value modeling but also reveals difficulties in dealing with the creation of 'information' entities like a schedule. These entities conform to two different types. They are a resource type and a schedule type. The solution proposed in the example utilized the typification relationship between the materialized schedule (resource) and the production schedule (schedule).

## 7 Discussion

The REA core pattern, depicted in Fig. 1 and 4, reveals the essence of REA ontology which is a paired 'transaction' with the duality and reciprocity relationships. Commitment and event entities are crucial entities in the REA value model. The duality and reciprocity relationships enable to define business event data in coherence at data level. On the other hand, it brings about the mutual binding of two different operations (give, take) which should be taken as one. However, above mentioned crucial entities miss vigorously defined states and state transitions. The current 'state' machine is imported from outside to comply with the relevant norms and standards. But the true state model should be firmly bound with the inner states of the commitment and event entities.

The REA core pattern and consequently REA value model is a kind of social system because two economic agents (human beings) play their roles in it. From this point of view, the REA core pattern (including commitment entities) should be in compliance with the DEMO *basic* and even *standard* and *universal* transaction pattern. The DEMO transaction pattern could provide necessary guidelines and vigorous definition of the mutual bonds between economic agents which also includes commitment and event entities and their states and state transitions. In addition, the compatibility with the DEMO transaction pattern might be also helpful for exception handling including revoking operations.

The fulfillment relationship that relates commitment to event entities should be semantically divided into smaller parts, in order to be unambiguous. The current solution is in a compliance with practice but it should be defined more precisely.

REA ontology doesn't explicitly provide any 'action model' and utilizes only the type entity concept for this reason. This concept has its utilization in specifications and guidelines for entities that conforms to a given type and for a 'category' specification of the planned entities such as a resource entity. Properly defined REA state machine could be a good starting point for systematic work on the REA business rule specifications and their tight integration in REA ontology.

In REA value model type entities are usually represented by resource type, event type and agent type entities. The function of event type entity is actually replaced by the commitment entity. The concept of 'actor role' in DEMO seems to be similar to the agent type in REA ontology. The only resource type entity still remains for further research.

REA value model used to be divided into two parts, the operational level that is roughly created by the REA core pattern and the policy level that roughly contains all the other entities. Both these levels should create a seamless unit but they don't. As an example, the difficulties in creating policy level entities such as contract and schedule may serve.

The benefit of REA domain ontology is that it enables to model business processes on the basis of assigning of the value of resources. The other important feature of the REA approach is that it operates on primary and raw economic data. For example, the quantity on hand for an inventory item can be calculated from the difference between the purchase and sale events or between the production and consumption events. It enables to offer a wider, more precise and more up-to-date range of reports. The REA business framework provides basis for designing accounting-based enterprise information systems from the scratch or semantically guided developments of existing ones. However, its weakness is the absence of vigorous theory and methodology. Particularly, it is not quite clear how to create entities such as contract and schedule. The other issue to solve are mentioned in the individual sections of the paper.

## 8 Conclusions and Further Work

The aim of this paper is to present REA domain ontology with the focus on its crucial areas. In some parts of the paper, an inspiration of DEMO (Design & Engineering Methodology for Organizations) was utilized. Although both approaches differ in their aim and usage (domain specific ontology & generic ontology), it was in any case worthwhile to analyze REA ontology with Enterprise Ontology in mind. There are four principal issues the future research should focus on.

- Applying the Performa-Informa-Forma analysis on all activities of REA value model.
- Introducing 'REA transaction pattern' for commitment and event entities similar to the DEMO's transaction pattern.
- Clarifying the REA type concept and put it in accordance with the ontological parallelogram.
- Utilizing fact based modeling perspective for REA value model.

REA ontology was developed and refined foremost by the needs of practice and that is why it is missing a vigorous theory and methodology. Utilizing DEMO can be beneficial and helpful to make REA ontology more transparent and unambiguously defined which increases its utilization possibly in accordance with DEMO.

**Acknowledgement.** The paper was supported by the grant provided by Ministry of Education, Youth and Sports Czech Republic, reference no. SGS08/PRF/2013.

## References

1. Gal, G., Geerts, Q., McCarthy, W.B.: The Ontological Foundation of REA Enterprise Information Systems. Presentation on VMBO Workshop Vien (2011), http://www.vmbo2011.ugent.be/VMBO2011/Program.html (cit. November 28, 2013)
2. Geerts, G.L., McCarthy, W.E.: The Ontological Foundations of REA Enterprise Information Systems. Paper presented at the Annual Meeting of the American Accounting Association, Philadelphia, PA (2000)
3. Geerts, G.L., McCarthy, W.E.: Policy-Level Specification in REA Enterprise Information Systems. Journal of Information Systems 20(2), 37–63 (2006)
4. Dietz, J.L.G.: Enterprise Ontology – Theory and Methodology. Springer (2006)
5. Dunn, C.L., Cherrington, O.J., Hollander, A.S.: Enterprise Information Systems: A Pattern Based Approach. McGraw-Hill/Irwin, New York (2004)
6. Horiuchi, S., McCarthy, B.: An Ontology-based State Machine for Catalog Orders presentation on VMBO Vien (2011), http://www.vmbo2011.ugent.be/VMBO2011/Program.html (cit. November 12, 2013)
7. Hruby, P.: Model-Driven Design Using Business Patterns. Springer, Heidelberg (2006)
8. Huemer, C., Zapletal, M.: A State Machine executing UMM Business Transactions. In: DEST 2007 Inaugural IEEE-IES, pp. 57–62 (2007)

9. ISO/IEC FDIS 15944-4: 2007. Business transaction scenarios - Accounting and economic ontology (2007)
10. Martin, J., Odel, J.J.: Object-Oriented Methods: A Foundation, UML Edition, 2nd edn. Prentice Hall (1997)
11. McCarthy, W.E.: The REA Accounting Model: A Generalized Framework for Accounting Systems in A Shared Data Environment. The Accounting Review, 554–578 (July 1982)
12. Vandenbosche, P.E.A., Wortmann, J.C.: Why accounting data models from research are not incorporated in ERP systems. In: 2nd International REA Technology Workshop, Fira, Santorini Island, Greece (2006), http://www.itu.dk/people/hessellund/REA2006/papers/Proceedings.pdf (cit. July 23, 2011)

# Evaluating Accounting Information Systems That Support Multiple GAAP Reporting Using Normalized Systems Theory

Els Vanhoof[1], Philip Huysmans[3], Walter Aerts[1,2], and Jan Verelst[3]

[1] Department of Accountancy and Finance, Faculty of Applied Economics,
University of Antwerp, Antwerp, Belgium
[2] School of Economics and Management, Tilburg University,
Tilburg, The Netherlands
[3] Department of Management Information Systems, Faculty of Applied Economics,
University of Antwerp, Antwerp, Belgium
{Els.Vanhoof,Philip.Huysmans,Walter.Aerts,Jan.Verelst}@uantwerp.be

**Abstract.** This paper uses a mixed methods approach of design science and case study research to evaluate structures of Accounting Information Systems (AIS) that report in multiple Generally Accepted Accounting Principles (GAAP), using Normalized Systems Theory (NST). To comply with regulation, many companies need to apply multiple GAAP. In case studies we identify AIS structures for multiple GAAP reporting. AIS need to cope with changes in GAAP and regulation in an evolvable way, the impact of the changes needs to be bounded. Since NST provides guidelines to design modular structures (in software) with an ex-ante proven degree of evolvability [1], we use NST to evaluate the identified AIS structures. We list violations of NST principles (combinatorial effects) and describe their manifestation in the cases. This application of NST in accounting demonstrates its relevance in non-software-specific domains. Moreover this is the first evaluation of an AIS with respect to evolvability.

**Keywords:** design science, Normalized Systems Theory, multiple GAAP, mixed methods.

## 1 Introduction

Many stakeholders, such as regulators, investors, customers, suppliers, or shareholders, need to be informed of the financial position of an organization. This presents a complex problem caused by the interplay of two separate issues: first, the financial position is not deterministic nor objective and second, accounting is handled by specialized Accounting Information Systems (AIS).

The first issue can be demonstrated by the existence of various Generally Accepted Accounting Principles (GAAP) which provide different instructions on how financial information should be reported. Different institutional environments and regulators can require financial information reports adhering to different GAAP [2]. Companies can be obliged to comply to the principles of multiple institutional environments and regulators because of several reasons. First of all, within one country the tax regulator might require the use of other principles than the filing agency, like for example in the Netherlands. A second reason can be that a company is listed on a European stock exchange: listed companies in the European Union are obliged to file their consolidated financial statements using International Financial Reporting Standards (IFRS). A third reason has to do with the international formation of corporate groups: when a company belongs to an international group, it might need to report its financial information to the holding company in the GAAP of the holding company, or the GAAP required by the stock exchange on which the holding company is listed, or both. For example, a Dutch company which is a subsidiary of a German holding company listed on both the Frankfurt as the New York stock exchange, might be required to report its financial information in Dutch GAAP, Dutch tax GAAP, IFRS, German GAAP and US GAAP. Because of these differences in regulatory requirements, some companies are obliged to process and report the same financial information using a different set of GAAP [2]. Moreover, a GAAP is not a static given and is subject to change. For example, recent events, like the credit crisis and corporate fraud scandals have caused an increasing demand of financial transparency (and enhanced quality and relevance) of company's financial information and thus have caused changes in accounting regulation.

The second issue is that in most modern organizations accounting is handled by specialized AIS, which are not always designed to handle reporting in multiple GAAP. Research suggests that information systems in the case of commercial off-the-shelf systems, are not easily customized or in the case of custom-built systems, become hard to adapt [3]. As a result, many organizations face difficulties when reporting in multiple GAAP.

Normalized Systems (NS) [4] has provided a theoretical response for the second issue of designing an AIS. It prescribes principles and design patterns to design information systems on which a set of anticipated changes can be easily applied. However, while the approach is generally applicable, the requirements for systems designed using NS need to be carefully formulated. For example, if redundant functionality is specified, changes to that functionality will still require large efforts. Therefore, NS does not completely solve the issue of AIS which need to report in multiple GAAP, as it provides no solution for handling the complexity of the differences in accounting rules.

In this paper we look into the problem that companies need information systems that support reporting in multiple GAAP in the changing regulatory environment. Compliance with regulation is a primary concern for every company and they prefer to comply in an efficient, but effective way. To do this, companies have set up structures in their AIS to be able to report in multiple GAAP, which we will evaluate in our evaluation section with respect to their evolvability, using NST.

We aim to contribute to a solution for this problem, by arguing that the problem of reporting in multiple GAAP can be interpreted as resembling the problems addressed by NS at the software level. Therefore indicating that this issue could be approached in a similar way as well. This rationale is the core of the design science methodology, which proposes to research novel solutions for wicked problems, based on solutions provided in related fields [5]. In Section 2, we elaborate on our use of design science methodology. Design science prescribes an iterative method of three phases: problem statement, design, and evaluation phase. The current paper reports on a first iteration through these phases. The problem statement phase describes the problem in more detail. This phase is the subject of Section 3. The design phase describes the design of proposed artifacts. In this paper, we focus on describing the artifacts which are currently used in real-life organizations. We describe our findings of this phase in Section 4. The evaluation phase evaluates the proposed artifacts according to a set of pre-defined criteria. We will demonstrate in Section 5 how we evaluated the described artifacts using criteria from NS theory. Finally, we will discuss this first iteration further in Section 6.

## 2 Methodology

In this research project we adopt the design science methodology, as our research interest concerns the utility of a design of a specific artifact for a given purpose. The use of design science in general is mainly motivated by the perceived lack of professional relevance of IS research [6,7]. This lack of relevance is addressed in design science studies by selecting a real-world problem and by designing an artifact that solves the problem or improves upon existing solutions. A variety of process models to conduct design science research are proposed in literature. All process models incorporate three main phases: a problem statement phase, a design phase, and an evaluation phase.

The relevance-oriented nature of design science is reflected in the words of Simon, who claims that "whereas natural sciences and social sciences try to understand reality, design science attempts to create things that serve human purposes" [8, p. 55]. The problem statement phase illustrates this intent by describing a real-world problem, and a justification of the value of a solution. The problem statement will determine the scope and evaluation criteria of a solution. Since the problems addressed in design science research are often hard or "wicked" problems, it follows that the problem statement should be clearly understood [9]. Therefore, it is considered useful to "atomize the problem conceptually so that the solution can capture its complexity" [10, p. 52]. In this paper, we elaborate on the problem statement of handling multiple GAAP in Section 3. In order to work towards a solution for this wicked problem, we need a structured understanding of the multiple GAAP problem. Therefore, we focus on the description of the design of currently-used artifacts for handling multiple GAAP reporting in practice. These designs are described and categorized in Section 4. We then select and apply a design science theory to provide a way of performing a rigorous evaluation of these empirical designs. This evaluation demonstrates whether and how the presented artifact designs provide a solution for the problem statement. The evaluation will be presented in Section 5.

This design iteration allows us to get a deeper understanding of the problem statement addressed in this research project, fulfilling the need to adequately capture and atomize the problem of handling multiple GAAP. Which can allow subsequent design iterations to be performed in a structured way.

Since the design iteration in this paper focuses heavily on understanding the designs of state-of-the-art artifacts currently used in practice, an empirical research component is deemed necessary. As a result, a set of case studies is conducted. This research design necessitates the adoption of a mixed methods framework, which allows the combination of design science and empirical methodologies.

### 2.1 Mixed Methods

In order to combine the strengths of different research methodologies, a mixed methods approach can be used [11,12]. In this paper the approach consists of two components (design science and case studies), studied separately, adhering to their respective methodology.

There are three main factors that influence the design of a mixed methods study. First, the theoretical drive of the project determines which component of the research is considered dominant in the context of the overall research project. In our project, this is the design science component, since the overall research goal is the evaluation of AIS artifacts. Second, the pacing of the components refers to how both components are synchronized in the context of the overall research project, concurrent or sequential. In our project, we have a concurrent pacing, since we perform qualitative case studies as a component of a design science iteration. Third, the point of interface refers to the stage at which the results from both components are combined or integrated. In our project, the design phase (of our design science component) is the point of interface, since we are studying the current design of AIS.

In conclusion, we can classify this research design as a *concurrent creative research design*, following the classification of [12]. This means that the qualitative component is enclosed in the design phase. The result of this phase will subsequently be used in the evaluation phase. Following the recommendations of the mixed methods methodology [11], we will explicitly discuss the methodology adhered to in the empirical research component in the next section.

### 2.2 Qualitative Method

In this paper we use case studies in the design phase. Using case studies is common in modularity research [13,14,15,16,17]. We use a multi-method approach to study multiple GAAP AIS in practice: next to our case studies we also interview two practitioners, experts in the field of AIS design in order to gain a deeper insight into the problems of multiple GAAP reporting.

The case studies that we conduct, are exploratory in nature, since the approach we use is novel: we apply existing knowledge (NST) on information systems designed specifically to support multiple GAAP reporting, which has not been done before [18,19]. Exploratory case studies are the appropriate approach since the phenomenon we study is contemporary; we investigate it in its natural setting to gain deeper insight into the matter [20,21] and our research question is centered around how AIS are designed to support multiple GAAP [21]. We choose a collective case study approach and a heterogeneous sample, because we want to be able to generalize our results: we want to identify combinatorial effects that are inherent in multiple GAAP AIS [22,19]. Hence we include companies from different sectors: an insurance company, a pharmaceutical manufacturer and two transportation firms.

The number of case studies is limited to four, since we need a high level of detail to be able to do the appropriate analysis [23,24,19]. This number falls within the range of four to ten cases, which is considered sufficient [25]. The population from which we draw our cases are all companies that report in multiple GAAP. Our first case study is chosen because we have contacts within the company through a colleague. To find additional cases we contacted Belgian subsidiaries of listed companies, because we know they have to report in multiple GAAP: they are obliged to file their statutory financial statements with the National Bank of Belgium using Belgian GAAP and their holding company is obliged to file their consolidated financial statements using IFRS (which includes the financial information of the subsidiary).

We use the multiple key informant method for data collection. Firstly, our informants need to be knowledgeable about the multiple GAAP design in their AIS and need to be willing to participate in our study [26,27]. Secondly, we conduct interviews with employees in the financial accounting department and in two cases also key informants from the IT department who are involved with the AIS. Interviewing multiple key informants provides us deeper insight into the specific design of the AIS.

The interviews are conducted by the first author of this paper and last approximately 120 minutes each. They all take place at the company site. Interviews are not recorded, rather notes are taken. These notes are electronically archived for later analysis. Our aim is to gain a profound insight into the design of the company's AIS. Therefore, we start interviews with a set of open questions to be able to ask detailed questions afterwards. Moreover, we do not use a fixed set of questions and adapt questions based on previous interviews. It is not unusual to alter and add data collection during the study in exploratory and theory-building research [25]. Since most of our contacts are financial accounting experts, we investigate the functional aspects of the system, rather than the technical aspects. We conduct one interview per participant, except with the financial accounting expert of the insurance company.

Next to our case studies, we interview two practitioners: someone from SAP Belgium who is responsible for localization (localization concerns country specific pre-configuration of SAP to comply with local regulation) of SAP and someone who works as a business intelligence consultant for a software company but permanently works with a client in the banking sector. These interviews help us to gain a deeper insight into the problems of multiple GAAP reporting, which allows us to analyze our case studies more thoroughly.

We analyze gathered data in two steps: within-case analysis and cross-case analysis. After each interview we analyze our notes to identify the structure used to set up a multiple GAAP AIS. Then we theoretically propose changes to the structure and evaluate the impact of the change on the existing structure, identifying combinatorial effects. We use a flexible and opportunistic data analysis [20,28,25,21]: when revealing a combinatorial effect in a certain case, we evaluate whether it also exists in previously analyzed cases. This is the first step in our cross-case analysis, after all cases are analyzed, we compare findings among cases, to be able to reanalyze cases and properly structure our findings. This results in the description of the different structures used to set up multiple GAAP AIS and the combinatorial effects resulting from different changes imposed on these structures. Lastly, we use insights from our expert interviews and online documentation of SAP to review and extend our analysis.

## 3 Problem Statement

Processing and reporting financial information in multiple GAAP requires knowledge of the differences between GAAP. There are five different ways in which GAAP can differ [29]:

- The definitions of concepts, for example, the difference between equity and liabilities.
- Recognition criteria: they determine if, when and how an item is recognized.
- Measurement methods: determining the amount included in the financial statements can be based on a different measurement method or model.
- Presentation: there are differences in the way financial statements should be presented regarding terminology, classification, which sections to use and the type of accounts to use.
- Disclosure: differences in the additional information to be included in the notes to the financial statements and the format/depth of these disclosures.

Moreover when reporting in multiple GAAP, two additional issues should be taken into account[29]:

- Alternatives: in many cases GAAP allow alternative recognition and measurement rules. For example, for measurement of inventory IFRS allows both the weighted average and the FIFO method.
- Lack of requirements or guidance: when one GAAP alternative does not address an issue that is specifically addressed in another GAAP alternative. IFRS 13 Fair Value Measurement for example, specifies how to perform fair value estimation, whereas in Belgian GAAP such guidelines do not exist.

This means we have to take all of these possible differences into account when we design AIS that support multiple GAAP. Compliance with additional GAAP is not a new issue, but information systems are not always set up in a way to make the implementation of an alternative GAAP easy: AIS are usually not built with the purpose of supporting multiple GAAP and this results in AIS that are not easy to maintain when regulation changes [30].

We consider an AIS, consisting of accounting components like a ledger, accounts, a chart of accounts, transactions and postings, as the modular structure in this paper. A ledger records the totals of postings to different accounts. The accounts are hierarchically structured in a chart of accounts (with the account name and number), which divides the accounts into logical categories (assets, liabilities, equity, revenue and expenses). When a transaction needs to be recorded in the financial statements, a posting is made to the ledger that consists of the accounts that increase and/or decrease with the respective amounts.

## 4 Design

In this section we describe different aspects of possible software designs for AIS that support multiple GAAP, based on our cases, our expert interviews and the SAP website [31]. We first describe some concepts by using the descriptions of SAP, later on we use these concepts to describe our case material.

A ledger contains all financial information presented according to one GAAP, using the accounts of the chart of accounts. A ledger is always contained within one company code, but one company code can contain different ledgers. Within a company code SAP requires all ledgers to use the same chart of accounts. To provide an overview of financial information of more than one company code, consolidation is necessary. [31]

All cases separate the accounts from different GAAP, so a posting is always made to the accounts of a specific GAAP. We identify three main ways to do this, namely the use of parallel accounts (accounts design 1), parallel ledgers (accounts design 2) or separate company codes (accounts design 3). To set up parallel accounts (accounts design 1) there are two different approaches: in one case all accounts are duplicated (accounts design 1a), in the other case different 'areas' of accounts (accounts design 1b) are defined: one joint area (for postings that are the same for all GAAP) and for every GAAP a separate area for postings. We visualize both cases in figure 1. One of our case companies uses the duplicated accounts design (accounts design 1a). They only have one chart of accounts that contains accounts twice: one for their primary GAAP, the other for their secondary GAAP. The description from the SAP website [31] for the areas design depicts that there is only one chart of accounts that contains common accounts and separate accounts for all additional GAAP. The areas design is not used by one of our case companies.

We do not have any cases that use parallel ledgers (accounts design 2), however this method is explained on the SAP website thoroughly. For every GAAP a separate ledger is created, based on the same chart of accounts, although a selection of accounts can be made [31].

Two of our case companies developed their own software package and use separate company codes (accounts design 3) to post to different GAAP. Although their configuration looks more like the parallel ledger (accounts design 2) option in SAP, since it does not require consolidation to add accounts from two GAAP. We will label this design as accounts design 3b. The company code option within SAP we will label as accounts design 3a.

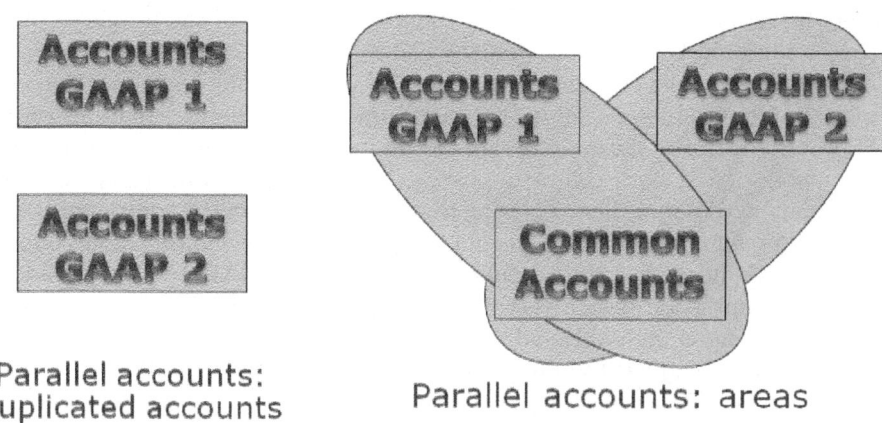

**Fig. 1.** Parallel accounts for two GAAP: design 1a and 1b

The design from our last case study, we will label accounts design 3c: they use different company codes for different GAAP, but also post operational transactions in a different company code than other transactions.

The second design choice concerning the setup of a multiple GAAP AIS is the choice whether to make difference postings (posting design 1) or complete postings (posting design 2). All our case companies use difference posting, although the insurance company has a special setup (posting design 3) regarding financial instruments. Difference posting (posting design 1) means that all postings are made to the accounts of the primary GAAP and only the difference between the primary and the additional GAAP is posted to the accounts of the additional GAAP. The use of this posting design results in the need to add the accounts of the primary and the additional GAAP to make reports for the additional GAAP. Complete posting (posting design 2) means that for every transaction a posting is made to both GAAP, even when the posting is the same for both GAAP. Posting design 3 results from the fact that the differences between GAAP concerning financial instruments are large and therefore the needed valuations for both GAAP are made independently. The insurance company therefore uses complete postings for financial instruments. But since for all other postings difference posting is used, they need to make an additional posting to the additional GAAP ledgers: a reverse posting of the posting to the primary ledger. When the ledgers are then added for reporting these postings cancel each other for the additional GAAP.

The use of difference postings in all cases has specific implications on the selection of accounts for reporting, depending on the accounts design. For duplicated parallel accounts (accounts design 1a) selection of accounts is needed when reporting: only the accounts of the primary GAAP or all accounts. The two cases that use accounts design 3b use the same chart of accounts in all company codes. For reporting in the primary GAAP they only need the primary company code, reporting in the additional GAAP requires adding all accounts for both company codes. In accounts design 3c consolidation of two company codes (operations and transactions) is required to report in the primary GAAP, for the additional GAAP four company codes should be consolidated.

For the implications of the posting design on accounts design 1b, accounts design 2 and accounts design 3a we base ourselves on the explanations on the SAP website [31]. For the areas design (accounts design 1b) and parallel accounts (accounts design 2) both complete postings and difference postings are allowed, and it is allowed to mix them. Causing that for reporting in the additional GAAP one should evaluate for every account separately whether only the account of that GAAP is needed or the account of the primary GAAP needs to be added. In the company code design (accounts design 3a) automatic postings are not possible, posting in an additional company code needs to be done manually and consolidation is required to add accounts from two company codes.

To conclude this section we make two additional observations. First, we observe that our case companies try to limit the differences between GAAP to a minimum: when they can choose between accounting treatments they will choose the one that allows them to make the same postings in all GAAP. The main differences between GAAP concern treatment of provisions, deferred taxes, pension benefits and financial instruments. Second, our four case companies all use a maximum of two different GAAP (Belgian GAAP and IFRS or another European GAAP), although with the future introduction of Solvency II the insurance company will need to produce a statement of financial position following the Solvency II directive, which can be considered as an additional GAAP.

## 5 Evaluation

The previous section demonstrates different designs which are used to implement the ability to report financial information under different GAAP. The lack of a single "best practice" or generally agreed upon design, and the remaining perceived complexity concerning multiple GAAP indicates that a satisfactory design has not yet surfaced. In order to be able to assess the different designs in a more structured way, and to identify requirements for a new design iteration, we propose an evaluation based on a design theory. This addresses the need for a rigor cycle in design science research projects [32]. We select the Normalized Systems Theory (NST) for several reasons.

First, NST is valid for our evaluation goal since it addresses *changing complexity* in systems. Originally, NST was applied to the software domain [1,4], but has later been applied to business processes and enterprise architecture domains [33,19], and has been described as a general design theory [34]. Moreover, it has already been applied in the cost accounting domain [35]. For our research, we focus on the complexity in AIS which needs to respond to changes as a result of handling reports in multiple GAAP. The changes which will be evaluated are:

- An additional account needs to be created.
- New revenue recognition criterion (effect on postings).
- New revenue recognition criterion (effect on modules).
- New measurement criteria for all GAAP.

Second, we select NST since it provides clear evaluation criteria, i.e., the occurrence of combinatorial effects. Combinatorial effects have been identified as the main obstacle for dealing with complexity in changing systems. A combinatorial effect occurs when the impact of a change in an information system not only depends on the size of the change itself, but also on the size of the system. As a simple example of a combinatorial effect, consider the use of a specific document to report travel expenses. If every employee keeps a stock of such documents for his or her own usage, a change in this document implies that the stock of each employee should be replaced. As a result, the impact to implement this change grows with the amount of employees (i.e., the size of the system). NST therefore prescribes that the evaluation should consist of the identification of combinatorial effects during the implementation of changes in the proposed designs. The identification of such combinatorial effects indicates the lack of a well-structured approach to handle these changes, which will lead to an increasing complexity in the artifact.

Third, the selection of NST provides us with a set of guidelines which indicate how the identified combinatorial effects can be prevented in subsequent design cycles. This enables future design iterations, aligned with the current, more empirically-oriented, design iteration. This alignment of design cycles provides an opportunity to contrast design proposals directly with the observed state of the art. This contrasting is an important part of the relevance cycle in design science research [32].

In the remainder of this section, we study how the anticipated changes could be implemented in the proposed designs.

**Change 1: An Additional Account Needs to be Created.** For example, because the company introduces a new product and wants to record income for that product on a separate account.

- **Context 1:** Impact in case of parallel accounts with duplicated accounts (account design 1a).
    - A new account needs to be created for all duplicated sets of accounts: the impact of the change is equal to the number of GAAP. Since the impact of the change is dependent on the size of the modular structure it is imposed on (the number of GAAP), this is a combinatorial effect.
    - The selection of accounts needed for reporting has to be changed for all GAAP, unless they are set up in a way to prevent this impact. For example, one of the attributes of an account depicts to which GAAP the account belongs. Since the impact of the change depends on the size of the modular structure it is imposed on (the number of GAAP), this is a combinatorial effect.
- **Context 2:** Impact in case of parallel accounts with three areas of accounts (account design 1b).
    - The impact depends on whether the revenue account needs to be added in:
        * **The common area:** only one new account needs to be created. No combinatorial effect exists.

        * **The GAAP specific areas:** the number of accounts that need to be created depends on the number of GAAP, which makes the change dependent on the size of the modular structure in place. Therefore, a combinatorial effect exists.
    • The selection of accounts needed for reporting in all GAAP has to be changed to include the newly created account(s), unless they are set up in a way to prevent this impact. For example, if one of the attributes of an account depicts whether the account is a common account, one for the primary GAAP or one for the secondary GAAP and so forth. If the accounts do not contain such an attribute, a combinatorial effects arises since the impact of the change depends on the number of GAAP used in the AIS.
  – **Context 3:** Impact in case of parallel ledgers (account design 2): a new account needs to be added to the chart of accounts that the ledgers use. Since the ledgers for the different GAAP all use the same chart of accounts the account needs to be added once and can be used by all ledgers. If not all accounts need to be used in all GAAP, one does need to specify upon creation of an account which GAAP will use it, which depends on the change. Hence no combinatorial effect arises from adding the new account.
  – **Context 4:** Impact in case of separate company codes (account design 3): the impact depends on whether the different company codes use the same chart of accounts or a different one.
    • When the same chart of accounts is used by all company codes (for example in account design 3b), the account only needs to be added once. Like in context 3, no combinatorial effect arises.
    • When a different chart of accounts is used, the addition of a new account causes an addition in all charts of accounts of all separate company codes. Since the impact of this change depends on the number of GAAP used in the systems, this is a combinatorial effect.

**Change 2: New Revenue Recognition Criterion (Effect on Postings):** One GAAP changes the revenue recognition criteria from when the sales contract is signed to when delivery of the goods takes place. Two separate situations can be distinguished.

a) In the past revenue recognition criteria are the same for all GAAP.

b) In the past the primary and at least one other GAAP had different revenue recognition criteria.

  – **Context 1:** Impact in case of parallel accounts with three areas of accounts (account design 1b):
    • For situation a: the revenue accounts in the common area should no longer be used and revenue accounts need to be added to all GAAP specific areas. Since the impact of the change depends on the number of GAAP in which reporting is needed, a combinatorial effect arises from changing the revenue recognition criteria. This combinatorial effect only exists in the case of parallel accounts with three areas of accounts (accounts design 1b), in other accounts designs (1a, 2 and 3) the new revenue

recognition criteria do not require adding new accounts to the chart of accounts.
  - For situation b: The accounts already exist in the GAAP-specific areas, so no additional change is needed. Hence, no combinatorial effect arises.
- **Context 2:** Impact in case of difference posting (posting design 1): the posting for the additional GAAP depends on the posting (accounting treatment) of the primary GAAP. The impact of the change is different depending on situation a or b as described above and on whether it is the primary GAAP or the additional GAAP that changes.
  - Situation a + an additional GAAP changes: a posting to the additional GAAP ledger needs to be created. There is no combinatorial effect.
  - Situation b + an additional GAAP changes: only the posting to the additional GAAP ledger changes. There is no combinatorial effect.
  - Situation a/b + the primary GAAP changes: the posting to the primary GAAP ledger needs to be changed, because all postings to other GAAP are difference postings depending on the posting to the primary GAAP ledger. This is a combinatorial effect, since the impact of the change depends on the number of GAAP in the AIS.
- **Context 3:** Impact in case of complete posting (posting design 2): postings to the different GAAP ledgers do not depend on each other. When recognition criteria change and the related postings need to be altered, they do not affect each other. For both situation a and b no combinatorial effect arises.
- **Context 4:** Impact in case of special setup (financial instruments) (posting design 3): postings to the other GAAP ledgers depend on the posting to the primary GAAP ledger in such a way that the reverse posting of the primary GAAP needs to be made as well. This is not a combinatorial effect, unless the reverse posting is specified individually in each posting to each GAAP. So when the reverse posting is separated from the posting to the individual GAAP ledger no combinatorial effect arises. This is true for both situations a and b.

**Change 3: New Revenue Recognition Criterion (Effect on Modules):** One GAAP changes the revenue recognition criteria from recognition when the sales contract is signed to when delivery of the goods takes place. Applies to all accounts designs and postings designs.

- **Context 1:** Impact when measurement is not separated into a different module: A new module needs to be created for the new recognition criterion, but since the measurement method is present within the same module, it needs to be duplicated into the new module. Two versions of the measurement method are created in this way. If in the future the measurement needs to be adjusted, a combinatorial effect arises because the impact then depends on the number of different combinations of recognition and measurement criteria.

- **Context 2:** Impact when measurement is separated into a different module: Creation of the new module for revenue recognition does not cause a new combinatorial effect.
- **Remark:** This is not only so for recognition and measurement, but for all ways in which GAAP can differ: definition of concepts, recognition criteria, measurement methods, presentation and disclosure.

**Change 4: New Measurement Criteria for All GAAP** : All GAAP use the same measurement method for example, fair value. But because of a change in market conditions the way the fair value of a certain category of financial assets is determined, needs to change. Applies to all accounts and postings designs.

- **Context 1:** When every GAAP has its own fair value calculation method (because for example, recognition and measurement are not split into separate modules) all the fair value calculation modules need to be changed as result of the changing fair value method. The impact of the change depends on the number of GAAP used in the AIS, hence a combinatorial effect arises.
- **Context 2:** When all GAAP use the same fair value calculation module, which is also separated from recognition, concepts, presentation and disclosure, the impact of the fair value calculation change is limited to the responsible module. There is no combinatorial effect.

# 6 Discussion, Future Work and Conclusion

By the use of a mixed methods approach of design science and case study research, we evaluate existing structures of AIS that report in multiple GAAP, with the use of NST. We find combinatorial effects in each of these structures, although the manifestation of these combinatorial effects differs. Hence we can conclude that some structures are more evolvable than others. For example, using parallel ledgers (accounts design 2) avoids the combinatorial effect that arises from using parallel accounts (accounts design 1).

This paper contributes to the literature in two ways. Firstly within the AIS literature this is the first time a system is evaluated with regard to evolvability. Secondly this is one of the first attempts to apply NST in a domain specific area [34,36]. By identifying combinatorial effects in multiple GAAP AIS, we show the applicability of NST in a non-software domain.

In future research NST could be applied in other domains for example logistics or production. Nevertheless within accounting more further research is possible: first of all the NS approach does not only identify combinatorial effects, but also proposes solutions to avoid them. In future work we will attempt to propose solutions in the form of guidelines that can be used when designing a multiple GAAP AIS. For example, a first guideline could depict that the same chart of accounts should be used by all GAAP.

# References

1. Mannaert, H., Verelst, J.: Normalized Systems: Re-creating Information Technology Based On Laws For Software Evolvability. Koppa (2009)
2. Sinnett, W.M., Willis, M.: The time is right for standard business reporting. Financial Executive 25(9), 23–27 (2009)
3. Lehman, M.: Programs, life cycles, and laws of software evolution. Proceedings of the IEEE 68, 1060–1076 (1980)
4. Mannaert, H., Verelst, J., Ven, K.: The transformation of requirements into software primitives: Studying evolvability based on systems theoretic stability. Science of Computer Programming 76(12), 1210–1222 (2011)
5. Hevner, A.R., March, S.T., Park, J., Ram, S.: Design science in information systems research. MIS Quarterly 28(1), 75–105 (2004)
6. Benbasat, I., Zmud, R.W.: Empirical research in information systems: The practice of relevance. MIS Quarterly 23(1), 3–16 (1999)
7. Hirschheim, R., Klein, H.K.: Crisis in the IS field? a critical reflection on the state of the discipline. Journal of the Association for Information Systems 4(5), 237–293 (2003)
8. Simon, H.A.: The Sciences of the Artificial, 2nd edn. The MIT Press (1969)
9. Buchanan, R.: Wicked problems in design thinking. Design Issues 8(2), 5–21 (1992)
10. Peffers, K., Tuunanen, T., Rothenberger, M., Chatterjee, S.: A design science research methodology for information systems research. Journal of Management Information Systems 24(3), 45–77 (2007)
11. Morse, J.M., Niehaus, L.: Mixed Method Design: Principles and Procedures. Developing Qualitative Inquiry, vol. 4. Left Coast, Walnut Creek (2009)
12. Huysmans, P., De Bruyn, P.: A mixed methods approach to combining behavioral and design research methods in information systems research. In: Proceedings of the 21st European Conference on Information Systems (2013)
13. Campagnolo, D., Camuffo, A.: The concept of modularity in management studies: A literature review. International Journal of Management Reviews 12(3), 259–283 (2010)
14. Djelic, M.L., Ainamo, A.: The coevolution of new organizational forms in the fashion industry: A historical and comparative study of france, italy, and the united states. Organization Science 10(5), 622–637 (1999)
15. Miozzo, M., Grimshaw, D.: Modularity and innovation in knowledge-intensive business services: It outsourcing in germany and the uk. Research Policy 34(9), 1419–1439 (2005)
16. Salvador, F., Forza, C., Rungtusanatham, M.: Modularity, product variety, production volume, and component sourcing: theorizing beyond generic prescriptions. Journal of Operations Management 20(5), 549–575 (2002)
17. Thyssen, J., Israelsen, P., Jørgensen, B.: Activity-based costing as a method for assessing the economics of modularization-a case study and beyond. International Journal of Production Economics 103(1), 252–270 (2006)
18. Marchall, C., Rossman, G.: Designing qualitative research, 3rd edn. Sage Publications (2006)
19. Huysmans, P.: On the Feasibility of Normalized Enterprises: Applying Normalized Systems Theory to the High-Level Design of Enterprises. PhD thesis, University of Antwerp (2011)
20. Benbasat, I., Goldstein, D.K., Mead, M.: The case research strategy in studies of information systems. MIS Quarterly 11(3), 369–386 (1987)

21. Yin, R.K.: Case study research: design and methods, 3rd/4th edn. Applied social research methods series, vol. 5. Sage Publications, Newbury (2003)
22. Ritchie, J., Lewis, J., Elam, G.: Designing and selecting samples. In: Qualitative Research Practice: A Guide for Social Science Students and Researchers, pp. 77–108. Sage Publications, Thousand Oaks (2003)
23. Stake, R.: Case Studies. In: Handbook of Qualitative Research, pp. 236–247. Sage Publications, Thousand Oaks (1994)
24. Stake, R.: Case Studies. In: Strategies of Qualitative Inquiry, pp. 86–109. Sage Publications, Thousand Oaks (1998)
25. Eisenhardt, K.M.: Building theories from case study research. The Academy of Management Review 14(4), 532–550 (1989)
26. Campbell, D.T.: The informant in quantitative research. American Journal of Sociology 60(4), 339–342 (1955)
27. Kumar, N., Stern, L.W., Anderson, J.C.: Conducting interorganizational research using key informants. The Academy of Management Journal 36(6), 1633–1651 (1993)
28. Dubé, L., Paré, G.: Rigor in information systems positivist case research: Current practices, trends, and recommendations. MIS Quarterly 27(4), 597–635 (2003)
29. Fischer, M., Marsh, T.: Accounting and reporting convergence. International Journal of the Academic Business World 6(1), 1–10 (2012)
30. Meall, L.: Can you comply? Accountancy 133(1329), 73–74 (2004)
31. SAP AG: Sap documentation (2013)
32. Hevner, A.R.: A three cycle view of design science research. Scandinavian Journal of Information Systems 19(2), 87–92 (2007)
33. Van Nuffel, D.: Towards Designing Modular and Evolvable Business Processes. Doctoral dissertation, University of Antwerp (2011)
34. Huysmans, P., Oorts, G., De Bruyn, P., Mannaert, H., Verelst, J.: Positioning the normalized systems theory in a design theory framework. In: Shishkov, B. (ed.) BMSD 2012. LNBIP, vol. 142, pp. 43–63. Springer, Heidelberg (2013)
35. Huysmans, P., De Bruyn, P.: Activity-based costing as a design science artifact. In: Proceedings of the 47th Hawaii International Conference on System Science (HICSS 2014), pp. 3667–3676 (2014)
36. Verelst, J., Silva, A.R., Mannaert, H., Ferreira, D.A., Huysmans, P.: Identifying combinatorial effects in requirements engineering. In: Proper, H.A., Aveiro, D., Gaaloul, K. (eds.) EEWC 2013. LNBIP, vol. 146, pp. 88–102. Springer, Heidelberg (2013)

# Modeling Financial Statement Preparation of a SME Enterprise by an Accountancy Firm

Joop de Jong

Mprise
P.O. Box 598, 3900 AN Veenendaal, The Netherlands
jdjong@mprise.nl

**Abstract.** Information requirements of business actors serve as basic specifications for the design of information processes. However, most enterprises in the SME domain lack the information process for annually preparing the financial statement. This is not uncommon because preparing the financial statement is usually done by an external independent accountant. Preparing the financial statement does not only mean carrying out a sequence of logical steps, but it also means taking decisions about several issues, for example, the current value of fixed assets, the amount of obsolete stock and doubtful accounts. This paper discusses the fulfillment of actor roles through the accountant. Some of these actor roles are defined in the demanding organization; others are defined in the accountancy organization. This paper exhibits the construction models of the business organization as well as of the infological organization from both organizations. It provides a clear view about the responsibilities of the accountancy firm towards the demanding enterprise. The models exhibit that the business organization of the accountancy firm does not include only ontological transactions but also infological transactions.

**Keywords:** enterprise engineering, information management, construction model, enterprise ontology, financial statement modeling, case study.

## 1 Introduction

The enterprise engineering paradigm understands an enterprise essentially as a social system, of which the elements are human beings in their role of social individuals, bestowed with appropriate authority and bearing the corresponding responsibility [1]. It is striking that almost all research in the field of enterprise engineering pertains to the business part of the enterprise. It is true that information systems are considered as supportive to individuals in the business organization, but these systems are not understood as social systems but as rational systems [2-10]. De Jong [11] discusses the information organization in the enterprise engineering field as a social system, but there are still open questions. One of them is the question about modeling the acts of an external accountant in preparing the financial statement of an enterprise. Once a year, after closing the financial year, the financial statement of the enterprise has to be prepared. The financial statement of an enterprise is a formal record of the financial activities of its business. It includes relevant financial information in a structured manner and in a form easy to understand. It typically includes basic financial statements, accompanied by a management discussion and analysis. The financial statement has to be prepared according to legal rules and according to professional guidelines of independent accountants. Although laws differ from country to country, approval of the financial statement is usually required for investments, financing, and tax purposes. The approval is usually given by independent accountants.

The key issue to be discussed in this paper is the notion of 'responsibility'. More specific, one discusses the responsibility of the enterprise on one side and the responsibility of the independent accountancy firm on the other side in preparing the financial statement which complies with legal rules

The paper is structured as following. Section 2 contains some important theoretical notions on which the research is grounded. The accountancy case is elaborated in detail in section 3. This section ends with construction models of both enterprises in which the actor roles of an accountant are modeled in particular. Some concluding remarks are presented in section 4.

## 2 Theoretical Notions

The research is grounded on three important axioms and the organization theorem which are taken from the ψ-theory [12]. Let us first give a short introduction to these axioms and theorem.

The first axiom states that the operation of the organization is constituted by the activities of actors, which are elementary chunks of authority and responsibility fulfilled by human beings. Actors perform two kinds of acts: production acts, P-acts for short, and coordination acts, C-acts for short. These acts have definite results, namely production facts, P-facts for short, and coordination facts, C-facts for short, respectively.

The second axiom states that coordination acts are performed as steps in universal patterns. These patterns, also called transactions, always involve two actor roles, i.e. two chunks of authority and responsibility. They are aimed at achieving a result, the P-fact. Figure 1 shows the standard transaction pattern. One of the two partaking actor roles is called the initiator, the other the executor of the transaction. The initiator and the executor seek to reach consensus about the P-fact that the executor is going to create as well as the intended time of the creation. Whenever this P-fact is created, the initiator and the executor seek to reach consensus about the P-fact that is produced as well as the actual time of the creation (both of which may differ from the request). Only if this agreement is reached will the P-fact become existent. The request->promise->execute->state->accept path in Figure 1 is called the basic pattern; it is the course that is considered when the initiator and the executor consent every time. However, they may also dissent. There are two states from which this may happen, namely the 'requested' and 'stated' states. Instead of promising, an actor may respond to a request by declining it, and instead of accepting, one may respond to a statement by rejecting it. These responses put the process in the 'declined' and 'rejected' states respectively. These states are indicated by a double disk in Figure 1, meaning that they are discussion states. If a transaction ends up in a discussion state, the two actors must 'sit together', discuss the situation at hand and negotiate how to get out of it. The possible outcomes are a renewed request or statement (probably with a modified proposition) or a failure (quit or stop).

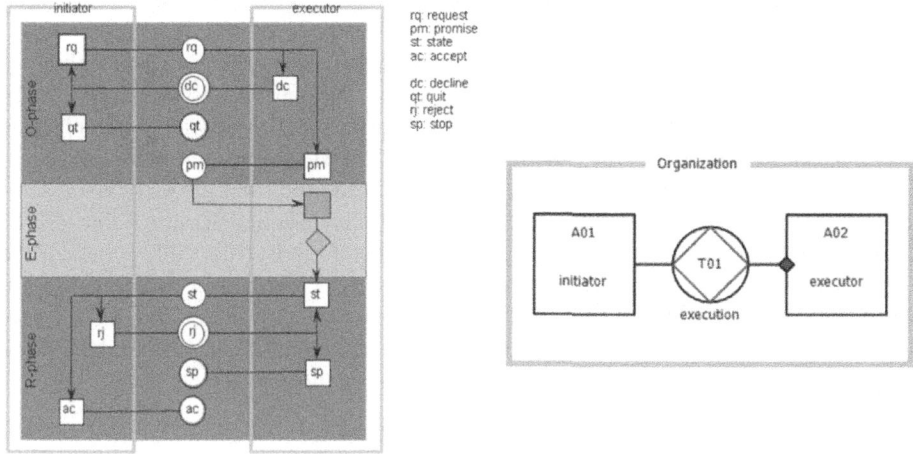

**Fig. 1.** The standard transaction pattern elaborated (left) [12] and in EE notation (right)

The third axiom states that there are three distinct human capabilities playing a role in the operation of actors, called performa, informa, and forma. These capabilities are recognized in both kinds of acts that actors perform. The informa capability is the human capability to carry out intellectual actions, such as reasoning, computing, remembering, and recalling thoughts. These are all actions by which content, irrespective of the form, is of value. Actors, which use the informa ability to perform P-acts, are called infological actors, I-actors for short. The forma capability is the human capability to carry out significational actions, such as transforming the form of content. Actors, which use the forma ability to perform P-acts, are called datalogical actors, D-actors for short. The performa ability is the human capability to carry out new, original actions, such as decisions and judgments. The performa ability is considered as an essential human capability to do business, of any kind. Actors, which use the performa ability to perform P-acts, are called ontological actors, O-actors for short.

The organization theorem states that the organization of an enterprise is a social system that is constituted as the layered integration of three homogeneous systems: the B-organization, the I-organization, and the D-organization. The business of an enterprise is brought about through the transaction kinds of which the initiator is an environmental actor role. These transaction kinds, their corresponding actor roles, as far as they are internal actor roles, are called the B-organization of the enterprise. D-actors in the D-organization support I-actors of the I-organization, while I-actors in de I-organization support B-actors of the B-organization. In a non-published paper, a B-actor could be an O-actor, an I-actor or a D-actor [13]. All three systems are called aspect systems of the overall organization of the enterprise. Integration between the three organizations is established through the cohesive unification of human beings. Let us elaborate this point in detail. However, how does an O-actor receive information from an I-actor? The answer is given by the distinction axiom. The subject, who fulfills the O-actor role, shapes into its informa ability for initiating an infological transaction with an I-actor in order to obtain the requested information [12, 14].

There are defined several aspects models of the business organization within the enterprise engineering field which are grounded on the mentioned theory [12]. Recently, some of these aspect models are also applied on the infological organization [11]. In this paper, we use the Interaction Model (IAM) and the Action Model (AM). The IAM of an organization consists of the transaction kinds and the recognized actor roles that participate as initiator or executor. The Transaction Result Table (TRT) defines the transaction kinds with the associated result kinds. This table is part of the IAM. The AM specifies the action rules that serve as guidelines for the actors in dealing with their agendas. The AM contains one or more action rules for every agendum kind. The AM is the most detailed and comprehensive aspect model. At the ontological level of abstraction, there is nothing below the AM.

## 3 The Accountancy Case

### 3.1 Information Services

Accountants are legally certified for preparing financial statements according to legal rules. The objective of a financial statement is to provide information about the financial position, performance and changes in the financial position of an enterprise that is useful to a wide range of users in making economic decisions. Financial statements should be understandable, relevant, reliable and comparable. Reported assets, liabilities, equity, income and expenses are directly related to an organization's financial position. Financial statements may be used for different purposes. Managers require a financial statement for making important business decisions that affect the continued operations of the enterprise. Prospective investors make use of a financial statement for assessing the viability of investing in a business and financial institutions (e.g. banks) use a financial statement for deciding whether to grant working capital or long term bank loans.

A financial statement can be prepared at any time. Usually, the financial statement is prepared by an accountant once a year, but events as merging and acquisition of businesses or gaining a long term bank loan require a more often preparation.

### 3.2 Accountability

The management of the enterprise has to realize them that the accountant needs access to all relevant original facts which are stored within the boundaries of the enterprise for executing the order. Based on these original facts the financial statement may be prepared by executing a number of infological steps sequentially. That presumes that all needed original facts have been created. This turns out to be in practice an incorrect assumption. We come back later on this issue.

The central question, which has to be answered now, is whether the experts of the accountancy firm fulfill actor roles in the organization of the enterprise, or do they fulfill actor roles in their own organization, or do they fulfill actor roles in both organizations.

**Table 1.** Organizational Construction Rules for splitting

| PC03 | … they cannot have a supporting role for other actors | The relevant actors prepare the financial statement in a joint effort. This effort results in a common report. The cooperation between the enterprise and the accountant is based on a business transaction between the enterprise and the accountant. By this transaction, the accountant is ordered to deliver the financial statement report and is authorized to use the original facts of the enterprise. Conclusion: the infological production acts in order to prepare the financial statement can be executed under the responsibility of the accountant. |
|---|---|---|
| PC05 | …. they operate under the same regulatory, legal and tax regime | The relevant actors prepare the financial statement according to the legal rules and professional guidelines. |
| PC08 | … they need comparable competences | The competences of the relevant actors needed to prepare the financial statement are comparable. For many middle-sized companies, the financial statement is prepared by one or two financial experts. |
| PC09 | … a (business) transaction relationship exists between them | In many cases, there exists a long term business relationship between the accountant and the enterprise. |
| PC10 | … an information-relationship exists between them | The original facts, which are needed for preparing the financial statement, are created in the enterprise and are owned by the enterprise. The financial statement must be understood as a derived fact. Derivations of facts are done by an accountant. |
| PC11 | … they have High Internal Cohesion and Low External Coupling | The accountant prepares the annual financial statement after finishing the year to which it relates. The complete set of relevant original facts has been frozen. This set has to be copied to the accountant so that he can derive the financial statement according to its own professional and legal rules. |

Some managerial handles about this issue are given by Op 't Land [15]. His research delivered eleven organization construction rules which lead to adequate splitting and allying enterprises. Although his research was primary focusing on splitting and allying the business part of enterprises, some of these rules are also applicable for the infological part of enterprises (cf. Table 1).

Table 1 shows that there are several good reasons for preparing the financial statement as an external service by an external accountant. Then, the accountant is fully responsible for the quality of deriving infological acts within the boundaries of its own company. Rule PC10 refers to an aberrant position because all relevant original facts have been stored in the enterprise. Then, insourcing of the accountant could also be an option. However, rule PC11 raises this objection.

Pertaining to the case to discuss, we propose an additional organization construction rule which does not belong to the set of Op 't Land. This rule is specifically applicable to the infological organization. It concerns the periodicity of the information process for deriving the financial statement. This information process is often not defined in an enterprise, due to the low periodicity of this process. On the other hand, the process of deriving financial statements is one of the core business processes of an accountancy firm. The needed actor roles are defined for it, and their employees are able to fulfill these roles. From this point of view, it is obvious that the financial statement has to be derived within the boundaries of the accountancy firm.

It is usual, after deriving the financial statement, to plan a date between the accountant and the shareholders of the enterprise. During this session, the financial statement is discussed extensively. Before finishing the meeting the accountant requests one of the representatives of the enterprise for signing the letter of confirmation of the financial statement. The first impression was that they asked someone of the enterprise to accept the work they did. Indeed, that is true, but they ask more. The letter of confirmation, which has to be signed by the representative of the enterprise, contains also the following sentence:

*"Finally, we recognize our responsibility for the financial statement. We confirm that we agree with your custom built financial statement for the current financial year."*

This sentence in the letter of confirmation begs the question how the "responsibility for the financial statement" should be understood. Does that mean that the enterprise is responsible for:

1. the correctness of the original facts needed for deriving the financial statement, or
2. the correctness of the process of deriving the financial statement, or
3. both 1) and 2)?

It is true that the enterprise is fully responsible for all facts it has created, but it is not true that the enterprise is responsible for the quality of deriving the financial statement whereas the deriving process is implemented within the boundary of the accountancy firm. An interview with the accountant revealed the opinion that the enterprise is fully responsible for the quality of the facts as well as the quality of the deriving process. That is remarkable. He answered literally:

*"This is because we have to deal with our own code of conduct and regulations relating to the work of accountants. If you (as an accountant) make a mistake, you can, for example, gain suspension or removal from the register of auditors. Therefore, we must be able to stand behind the report with the financial statement (e.g. valuations and accounting policies)."*

Does he move responsibilities from its own company to the enterprise, or is there anything else that makes that the accountant takes this stand? Indeed, there is something else going on. That is clearly illustrated in the last sentence, the part between brackets. It appears that the accountant cannot be satisfied with only infological production acts for the preparation of the financial statement. He performs also ontological production acts within the organization boundary of the enterprise. Although the management of the enterprise is not aware of these ontological production acts, performed by an accountant, they are still fully responsible for these acts.

The financial statement must be understood as a derived fact. However, in practice all needed facts are not available. Even the most well organized enterprise lacks some facts which are needed for the financial statement. These facts have not been created because of a lack of applicability. Most of them are only needed for the preparation of the financial statement. An interview with an accountant revealed nine ontological production acts which are executed by them during preparation of the financial statement. It concerns the following production acts:

1. Assessing the current value of the tangible assets. Estimating the durability and the residual value.
2. Inventory of office stock. Identification and assessment of obsolete inventories.
3. Estimating the amount of the provision for doubtful accounts
4. Estimating the progress of ongoing projects, estimating any provision for loss-making projects
5. Determining the level of provisions
6. Estimating payables whose height is currently not known
7. Determining correctness of mutual fees
8. Determining whether non-balance sheet liabilities in the financial statements are disclosed
9. Checking whether relevant events are reported after the balance sheet date

All these acts have to be performed according to the professional rules of the accountant. In practice, these acts are performed by the accountant himself. He fulfills the ontological actor roles within the boundary of the enterprise and, thereby, he creates the needed original facts. From now, preparing the financial statement has to be understood as an infological process.

The next two sections contain discussions about the IAM of the enterprise for which the accountant prepares the financial statement and the IAM of the accountancy firm for which the mentioned accountant works.

## 3.3 Modeling the Enterprise

In this section, we discuss the integral IAM of the B- and I- organization of the enterprise for which the financial statement has to prepared (cf. Fig. 2). Whereas the construction of every enterprise differs, the uniqueness of the enterprise is presented by the composite actor O-CA03.

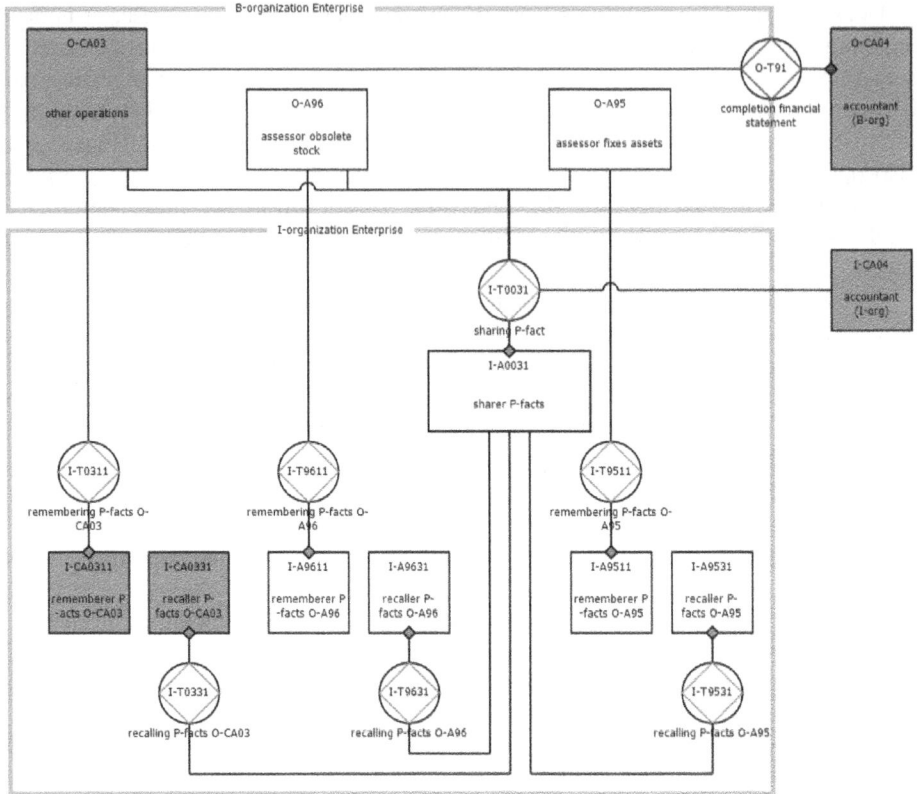

**Fig. 2.** The IAM of the enterprise for which the financial statement is prepared

Further, the B-organization contains two actor roles O-A95 and O-A96 which correspondents with the first two ontological production acts mentioned in the previous section (the first and the second act from the list of nine). The other ontological production acts from the list are not presented in the model due to the reason of keeping the model simple. Although these actor roles are fulfilled by experts of the accountancy firm, the management of the enterprise remains responsible for all ontological transactions within the boundary of the enterprise.

All facts they create are remembered by actors in the I-organization (I-A9511 and I-A9611) and are made available (I-A9531 and I-A9631) to the "sharer P-facts", I-A0031. All actors who want to use P-facts have to request I-A0031 for them. See for an extended description of the way of modeling of the I-organization in the thesis of De Jong [11]. Whereas the accountant prepares the financial statement within the boundary of his own organization, he (I-CA04) also initiates the transaction I-T0031 to get all original facts for deriving the financial statement. The accountant is authorized for this infological request by O-T91. Figure 3 exhibits that I-CA04 is constituted of the elementary actor roles I-A21 and I-A22.

## 3.4 Modeling the Accountancy Firm

This section discusses the integral IAM of the B- and I- organization of the accountancy firm (cf. Tab. 2 and Fig. 3).

**Table 2.** Transaction Result Table

| Transaction Kind | Result Kind |
|---|---|
| O-T91 – completion financial statement | O-R91: Fin.stat. [Fin_Stat] has been completed |
| I-T20 – preparation financial statement | I-R20: Fin.stat. [X] has been prepared |
| I-T21 – derivation fin.stat. part I | I-R21: Fin.stat. part I [Y1] has been derived |
| I-T22 – derivation fin.stat. part II | I-R22: Fin.stat. part II [Y2] has been derived |

Note: X, Y1 and Y2 are some information delivery.

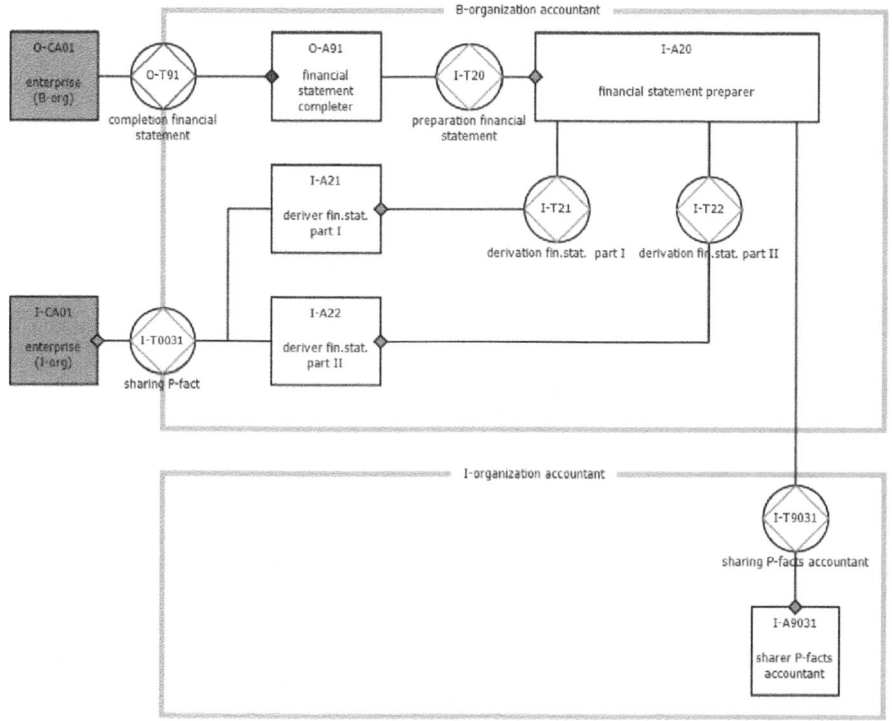

**Fig. 3.** The IAM of the accountancy firm

Figure 3 exhibits a new element which is so far not presented in a research paper; the B-organization includes not only ontological actor roles but also infological actor roles. The idea is based on a non-published paper of Dietz [13]. Both kinds of actor roles could bring about the business of an enterprise. That is certainly the case as fact processing is part of the business. The actor roles I-A21 and I-A22 request I-CA01 for getting the original facts of the enterprise, so I-T0031 initiates I-A0031 in the

I-organization of the enterprise. We mentioned in the previous section that these actors were authorized for initiating by the transaction O-T91. Whereas the financial statement is prepared after finishing the financial year, the entire set of needed original facts can be exported to the accountancy firm.

Deriving the financial statement has to be done according to legal rules en professional guidelines. These facts have to be invoked by the I-organization. Figure 3 exhibits that in the accountancy firm the transaction "sharing P-facts accountant", I-T9031, is initiated for gaining these facts.

Figure 4 exhibits a part of the AM of the accountancy firm. It specifies the action rules that serve as guidelines for the actors O-A91, I-A20, I-A21 and I-A22 in dealing with their agendas.

*Actor O-A91:*

| | | | |
|---|---|---|---|
| when | completion fin.stat. **for** new Fin_Stat is requested | | O-T91/rq |
| | with     **the** customer **of** Fin_Stat **is a** Customer | | |
| | **the** fiscal.year **of** Fin.Stat **is a** Year | | |
| assess | *justice:*    **the** Performer **of the** request **is the** customer **of** Fin_Stat | | |
| | *sincerity:*    <no specific condition> | | |
| | *truth:*       <no specific condition> | | |
| if | *complying with request is considered justifiable* | | |
| then | promise    completion fin.stat. **for** Fin_Stat | | |
| else | decline     completion fin.stat. **for** Fin_Stat | | |
| when | completion fin.stat. **for** Fin_Stat is promised | | O-T91/pm |
| assess | *justice:*    **the** Performer **of the** promise **is the** completer **of** Fin_Stat | | |
| | *sincerity:*<no specific condition> | | |
| | *truth:*       <no specific condition> | | |
| if | *complying with promise is considered justifiable* | | |
| then | request    preparation fin. stat. **for** X | | |

| | | | |
|---|---|---|---|
| when | preparation fin.stat. **for** X is stated | | I-T20/st |
| assess | *justice:*    **the** Performer **of the** state **is the** preparer fin. stat. **of** X | | |
| | *sincerity:*<no specific condition> | | |
| | *truth:*       <preparing financial statement is well done> | | |
| if | *complying with statement is considered justifiable* | | |
| then | | | |
| | accept preparing fin.stat. **for** X | | |
| | execute preparing fin. stat. **for** X | | |
| | state completion fin.stat. **for** Fin.Stat | | |
| else | reject     preparing financial statement **for** X | | |

# Modeling Financial Statement Preparation of a SME Enterprise

*Actor I-A20:*

| | | | |
|---|---|---|---|
| **when** | preparation fin. stat. **for** new X <u>is requested</u> | | I-T20/rq |
| **with** | **the** customer **of** X **is a** Fin.Stat.Customer<br>**the** fiscal.year **of** X **is a** Fin.Stat.Fiscal.year | | |

| | | |
|---|---|---|
| **assess** | *justice:* | **the** Performer **of the** <u>request</u> **is the** completer fin.stat. **of** Fin_Stat |
| | *sincerity:* | <no specific condition> |
| | *truth:* | <available fiscal rules><br><available professional guidelines> |

| | |
|---|---|
| **if** | *complying with request is considered justifiable* |
| **then** | <u>promise</u>  preparation fin.stat. **for** X |
| **else** | <u>decline</u>  preparation fin.stat. **for** X |

| | | |
|---|---|---|
| **when** | preparation fin.stat. **for** X <u>is promised</u> | I-T20/pm |

| | | |
|---|---|---|
| **assess** | *justice:* | **the** Performer **of the** <u>promise</u> **is the** preparer fin.stat. **of** X |
| | *sincerity:* | <no specific condition> |
| | *truth:* | <no specific condition> |

| | |
|---|---|
| **if** | *complying with promise is considered justifiable* |
| **then** | <u>request</u> derivation fin.stat. part I **for** Y1<br><u>request</u> derivation fin.stat. part II **for** Y2 |

| | | |
|---|---|---|
| **when** | derivation fin.stat part I **for** Y1 <u>is stated</u> | I-T21/st |

| | | |
|---|---|---|
| **assess** | *justice:* | **the** Performer **of the** <u>state</u> **is the** deriver fin.stat. part I **of** Y1 |
| | *sincerity:* | <no specific condition> |
| | *truth:* | <no specific condition> |

| | | |
|---|---|---|
| **if** | *complying with statement is considered justifiable* | |
| **then** | <u>accept</u> | derivation fin.stat part I **for** Y1 |
| **else** | <u>reject</u> | derivation fin.stat. part I **for** Y1 |

| | | |
|---|---|---|
| **when** | derivation fin.stat. part II **for** Y2 <u>is stated</u> | I-T22/st |

| | | |
|---|---|---|
| **assess** | *justice:* | **the** Performer **of the** <u>state</u> **is the** deriver fin.stat. part II **of** Y2 |
| | *sincerity:* | <no specific condition> |
| | *truth:* | <no specific condition> |

| | | |
|---|---|---|
| **if** | *complying with statement is considered justifiable* | |
| **then** | <u>accept</u> | derivation fin.stat. part II **for** Y2 |
| **else** | <u>reject</u> | derivation fin.stat. part II **for** Y2 |

| T20/ac | when | derivation fin.stat. part I **for** Y2 is accepted **and** derivation fin.stat. part II **for** Y2 is accepted | I- |
|---|---|---|---|
| | assess | *justice:* **the** Performer **of the** accept **is the** preparer fin.stat. **of** X<br>*sincerity:* <no specific condition><br>*truth:* <no specific condition> | |
| | if<br>then | *complying with acceptance is considered justifiable*<br>execute preparer fin.stat. **for** X<br>state preparation fin.stat. **for** X | |

*Actor I-A21:*

| | when | derivation fin.stat. part I **for** Y1 is requested<br>**with** **the** customer **of** Y1 **is a** X.Customer<br>**the** fiscal.year **of** Y1 **is a** X.Fiscal.year | I-T21/rq |
|---|---|---|---|
| | assess | *justice:* **the** Performer **of the** request **is the** preparer fin.stat. **of** X<br>*sincerity:* <no specific condition><br>*truth:* <no specific condition> | |
| | if<br>then<br>else | *complying with request is considered justifiable*<br>promise derivation fin.stat. part I **for** Y1<br>decline derivation fin.stat. part I **for** Y1 | |

| | when | derivation fin.stat. par I **for** Y1 is promised | I-T21/pm |
|---|---|---|---|
| | assess | *justice:* **the** Performer **of the** promise **is the** deriver fin.stat. part I **of** Y1<br>*sincerity:* <no specific condition><br>*truth:* <no specific condition> | |
| | if<br>then | *complying with promise is considered justifiable*<br>execute derivation fin. stat. part I **for** Y1<br>state derivation fin.stat. part I **for** Y1 | |

*Actor I-A22:*

| | when | derivation fin.stat. part II **for** Y2 is requested<br>**with** **the** customer **of** Y2 **is a** X.Customer<br>**the** fiscal.year **of** Y2 **is a** X.Fiscal.year | I-T22/rq |
|---|---|---|---|
| | assess | *justice:* **the** Performer **of the** request **is the** preparer fin.stat. part II **of** X<br>*sincerity:* <no specific condition><br>*truth:* <no specific condition> | |

| | | |
|---|---|---|
| **if** | *complying with request is considered justifiable* | |
| **then** | promise derivation fin.stat. part II **for** Y2 | |
| **else** | decline derivation fin.stat. part II **for** Y2 | |

| | | | |
|---|---|---|---|
| **when** | derivation fin.stat. par II **for** Y2 is promised | | I-T22/pm |
| **assess** | *justice:* | the Performer **of the** promise **is the** deriver fin.stat. part II **of** Y2 | |
| | *sincerity:*<no specific condition> | | |
| | *truth:* <no specific condition> | | |
| **if** | *complying with promise is considered justifiable* | | |
| **then** | execute derivation fin.stat. part II **for** Y2 | | |
| | state derivation fin.stat. part II **for** Y2 | | |

Fig. 4. The AM of the accountancy firm

## 4 Conclusion

A profound analysis shows that the accountant performs both ontological and infological transactions. These transactions are not discerned clearly from the perspective of responsibility. The ontological transactions are performed under the responsibility of the enterprise for which the financial statement is prepared, and the infological transactions are performed under the responsibility of the accountancy firm. A letter of confirmation of the financial statement gives rise to ambiguities. The accountant should only ask the management of the enterprise to sign for taking over the responsibility for the ontological production acts they execute. The accountant remains responsible for the quality of the infological production acts, in other words, the process of preparing the financial statement.

## References

1. Dietz, J.L.G.: Enterprise Engineering - the Manifesto. In: EEWC 2011, Antwerp, Belgium. Springer, Heidelberg (2011)
2. Mallens, P.J.M., Dietz, J.L.G., Hommes, B.J.: The Value of Business Process Modeling with DEMO prior to Information Systems Modeling with UML. In: Proceedings of the 6th CAISE/IFIP8.1 International Workshop on Evaluation of Modeling Methods in Systems Analysis and Design, EMMSAD 2001. Interlaken (2001)
3. Maij, E., et al.: Use cases and DEMO: aligning functional features of ICT-infrastructure to business processes. International Journal of Medical Informatics 65, 179–191 (2002)
4. Dietz, J.L.G.: Deriving Use Cases from Business Process Models. In: Song, I.-Y., Liddle, S.W., Ling, T.-W., Scheuermann, P. (eds.) ER 2003. LNCS, vol. 2813, pp. 131–143. Springer, Heidelberg (2003)

5. Shishkov, B., Dietz, J.L.G.: Deriving use cases from business processes: the advantages of DEMO. In: The Fifth International Conference on Enterprise Information Systems, Angers, France (2003)
6. Albani, A.: The Benefit of Enterprise Ontology in Identifying Business Components. In: 19th IFIP World Computer Congress, Santiago de Chile, Chile. Springer (2006)
7. Mulder, J.B.F.: Rapid Enterprise Design. Technical University Delft, Delft (2006)
8. Mannaert, H., Verelst, J.: Normalized Systems - Re-creating Information Technology Based on Laws for Software Evolvability. Koppa, Kermt (2009)
9. Terlouw, L.: Modularization and Specification of Service-Oriented Systems. Technical University Delft, Delft (2011)
10. Kervel van, S.J.H.: Ontology driven Enterprise Information Systems Engineering. Delft University of Technology, Tilburg (2012)
11. Jong de, J.: A Method for Enterprise Ontology based Design of Enterprise Information Systems. Technical University Delft, Veenendaal (2013)
12. Dietz, J.L.G.: Enterprise Ontology – theory and methodology. Springer (2006)
13. Dietz, J.L.G.: DEMO-3, Models and Representations, version 3.6c (March 2013), http://www.ee-institute.com
14. Dietz, J.L.G.: Architecture, building Strategy into Design. Academic Service, The Hague (2008)
15. Op 't Land, M., Applying Architecture and Ontology to the Splitting and Allying of Enterprises. Delft University of Technology, Delft (2008)

# Linking Value Chains – Combining e3Value and DEMO for Specifying Value Networks

João Pombinho[1,2], José Tribolet[1,2], and David Aveiro[1,3]

[1] INESC-ID and CODE - Center for Organizational Design & Engineering, INOV,
Rua Alves Redol 9, Lisbon, Portugal
[2] Department of Information Systems and Computer Science, Instituto Superior Técnico
University of Lisbon, Portugal
[3] Exact Sciences and Engineering Centre, University of Madeira, Funchal, Madeira, Portugal
`jpmp@ist.utl.pt, jose.tribolet@inesc.pt, daveiro@uma.pt`

**Abstract.** In this paper we provide a model for the bonding of systems in a value network. Our main contributions are: 1) a structural model of the chains and their viewpoints, and 2) a specification of how to use that structure within a process that supports the formalization of the rationale behind system development decisions. To provide a solution to this challenge we combine System Development and Value Modeling disciplines. From DEMO, we use the Generic System Development Process from the Tao-theory and its Value-oriented System Development Process implementation. We formalize basic concepts from e3Value, namely start stimulus, end stimulus, gates and scenario paths in an integrated way with system construction models. We provide a methodology for constructing e3Value models systematically and improve DEMO modeling by devising individual value networks in an adequate way and how different system components combine to form them.

**Keywords:** Value Network, Value Chain, Business Modelling, e3Value, DEMO, Enterprise Engineering.

## 1 Introduction and Motivation

The need of consistently and coherently design organizations is an increasingly pressing issue to modern enterprises. In fact, both academia and industry seem to have definitively embraced the topic, and rightfully so. Laudon notes that enterprise performance *is optimized when both technology and the organization mutually adjust to one another until a satisfactory fit is obtained* [2]. However, studies indicate as much as 90 percent of organizations fail to succeed in applying their strategies [3].

Misalignments between the *business* and its *support systems* are frequently appointed as a reason of these failures [2, 4]. Aligning Business and IT is a widely known challenge in enterprises as the developer of a system is mostly concerned with its function and construction, while its sponsor is concerned about its purpose, i.e., the system's contribution. Also, business vision of a system and its implementation by supporting systems is not modelled in a way that adequately supports the development and evolution of a system and its positioning in a value network.

Academic interest in Business Modeling has been increasing, as is the maturity of its practical application. The specification of Value Models brings to fore all the relevant actors involved in the interaction with the environment, as well as the significant value transactions and relations between them to provide value to systems in the environment. However, the issue remains how to integrate Business Modeling with constructional modeling in a way that allows designing and managing alignment.

In order to meet this challenge, it is necessary to relate the key concepts of the two disciplines, and create mechanisms to validate coherence that convey the conceptual relations specified, combining both disciplines in a system development process that supports rationale specification.

This paper presents an ontology for modelling value networks, supported by an integration between DEMO [5], from the perspective of Enterprise Engineering, and e3Value [6], from Value Modelling. The objective in matching DEMO and e3Value is introducing value-oriented reasoning for obtaining value-traceable outcomes in system design. We can summarize the mutual benefits as 1) being able to express the value context of any transaction as a manifestation of purpose; and 2) trace value-production to coordination/production facts/acts level, enabling system construction modelling. Using the ontology for inductive building of the value network, the approach consists in performing a conceptual system development process instantiation for consecutive nodes of the value network, using DEMO's *function/construction* alternation [7].

The paper begins by establishing the problem at hand in terms of a duality between purpose and construction and objectively defining the problem. In section 3, the matching ontology between DEMO and e3Value is presented. Following, in section 4 we present the Value Network Ontology, the main contribution, and the paper closes with conclusions and future work.

## 2 The Problem at Hand

The purpose of an organization is to create value, which is done by producing value objects towards its environment (the network). Organizations can be seen as artificial systems [5, 8, 9], as social systems [5] and also as actors in value networks [6]. Every artificial system, e.g., purposefully build by man, is a value system. Purposefully is the key term, as it conveys the intention that the creator has of extracting some kind of benefit from it. We hypothesize that it is both useful and feasible to create an integrated model that combines Enterprise Engineering and Value Modelling aspects while keeping coherence and formality between two different kinds of models not normally easy to match in a systematic and rational way.

## 2.1 Purpose and Construction

The purpose and the construction of a system are distinct notions, the first addressing its contribution towards the system's environment and the second addressing how the system brings about the said contribution. These concepts respectively relate to two aspects of a system: teleological, concerning its function and behaviour, a black-box; and ontological, about its construction and operation, a white-box [10]. Approaches in ICT development come mainly from the ontological side, as building the system is a necessary and defining contribution from such area.

A formal business model [11] is necessary for a grounded specification of constructional models. For instance, Business Model Ontology [12] contrasts with predominantly constructional approaches, such as Archimate [13]. The mapping between the two perspectives is not a novel idea [14] and rightfully so, as they are complementary and used at different stages of the development of a given system.

A method for value-oriented system development has been proposed [7], providing a specific context for the ontology specification here presented. In this sense, as it is used as a means to pursue teleological and ontological integration in a system development context, it fundamentally differs from other approaches, such as [15].

## 2.2 Problem Statement

Every artificial system is a value network. The relations that form between different elements (internal and external bonds) are motivated by the exchange of value objects. These value object exchanges are designed to happen in order to produce a certain end result (solution) for an actor. This actor has the initiative in transactions, or start stimulus of the solution chain. In turn, a chain holds together because of the equilibrium brought by forces in opposing directions. Analogously, a value chain holds together when its nodes are in equilibrium, i.e., they are economically viable, considering all the forces (i.e. value threads) passing through them. How to model this condition is also a primary challenge.

A solution for these issues must support the creation of multiple value chain perspectives and model how they are related in a holistic model. This essential, birds-eye view of the system is instrumental in managing complexity of the intertwined needs and motivations of stakeholders. Yet, at the same time, such a model must be formal in its specification and validation, and must be traceable to the remainder models specifying the detailed construction of the system.

In summary, the fundamental questions we address in this research are:

- How to model the structure of value as a combination of individual value chains, namely multiple threads joining in a single node?
- How to explain the origin of each value object and specify the role of each VO both as problem and as solution in a certain value thread?
- How to provide an objective approach to specify value feasibility?
- How to objectively specify the value rationale for all relevant components of organization models?

## 3 Matching e3Value and DEMO

In this section we present a brief introduction to a formal matching ontology with a set of rules, allowing bridging value exchanges and transactional modelling with the objective of bridging Value Modelling and Transactional Modelling. For illustration purposes, we will use the classic Library case from Enterprise Ontology [5]. In this example, two new streams that are implicit in the constructional model, one downstream from the Library and one upstream, are introduced. The first deals with mitigating the risk of non-return through a mechanism of membership that assures contracting and a dramatic increase in the collection capability. The second stream deals with the specifics of maintaining availability on a book title catalogue that is adequate to the demand.

e3Value is an ontological approach for modelling networked value constellations. It is directed towards e-commerce and supports analysing the creation, exchange and consumption of economically valuable objects in a multi-actor network.

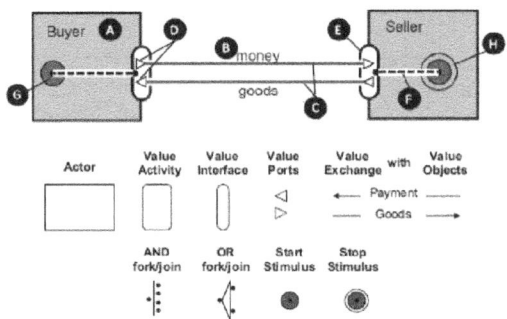

**Fig. 1.** Basic e3Value example and language components

In e3Value, an Actor (A, cf. Fig. 1) is perceived by his or her environment as an economically independent entity. They exchange Value Objects (B), which are modelled in combination with a Value Transfer (C); this transfer occurs through a Value Port (D), which is a directional element of a Value Interface (E). A Scenario Path (G, F, H)) represents a particular chain. The e3Value ontology specifying these concepts and their relationships can be consulted in detail in [6]. In [16], a formal matching ontology that unambiguously relates e3Value and DEMO models is presented. It comprises an objective set of concepts for integrating the teleological and ontological perspectives of a system (Fig. 2).

DEMO contributes to the constructability of a business model, by providing a theory and method for designing and engineering enterprises. e3Value contributes to the justifiability of a given system construction, providing the notion of purpose through value semantics, network context and economic concepts, namely reciprocity.

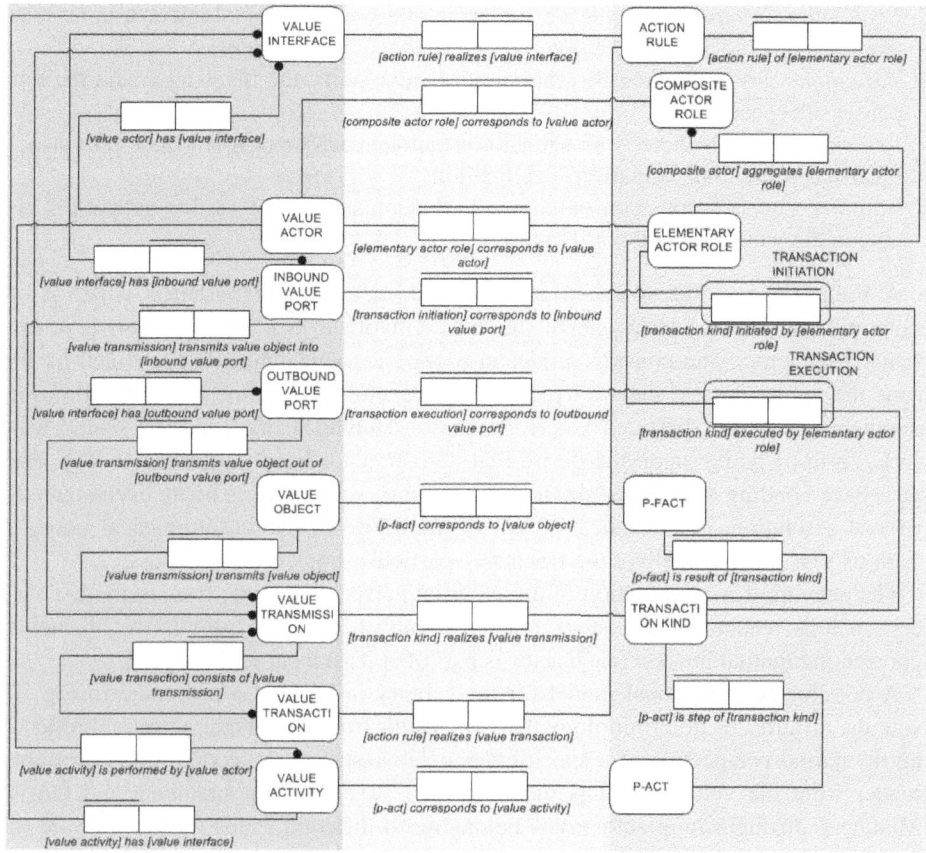

**Fig. 2.** Matching Ontology – ORM Model

Using the concepts from the matching ontology, a set of rules can be specified in order to be able to identify the set of misalignment points between an e3Value Model and a DEMO ontological model. The following rules, that allow checking the following integrated model qualities, were also defined:

1. *Value Object completeness* – for each actor (node in a VN), if a VO is exchanged in outbound value exchange, then it must either be exchanged in inbound value exchange or be generated inside, by a value activity; for instance, in the Library example, is the Book value object traceable from the final customer to its origin?
2. *Transaction completeness* – the acts and facts of the complete transactional pattern are fully specified; for instance, what happens if books are not returned, i.e., no *state* of the return book transaction is made?
3. *Value exchange constructability* – for each value exchange, the corresponding ontological transaction 1) exists and 2) is fully specified (2);
4. *Transaction justifiability* – for each ontological transaction, the corresponding value exchange 1) exists and 2) is complete (1);

5. *Economical reciprocity* – the dependencies between the ontological transactions that constitute a value transaction are specified via Action Rules;
6. *Valuation* – for each actor, the valuation of value ports, the Investment and the Expenses are specified;
7. *Implementability* – there is an actor to instantiate each actor role (value port structure) according to its valuation specification;
8. *Viability* – every value actor belonging to the composition of a value network is viable in the value network configuration.

A Value Object (VO) is a service, a product, or even an experience, which is of economic value for at least one of the actors involved in a value model. It is exchanged by actors who consider it has an economic value. The VO is defined by its name. Example: money, electricity, mp3, advice, etc. VOs define the motivation for establishing relations between actors, thereby assuming a primary role. Value networks form by acting upon the desire that a given actor has of a particular VO. The successive bonding between actors forms a chain of value objects being exchanged as parts of a solution. For instance, to satisfy the desire of reading a physical book, a chain of VOs forms to deliver the rights to a particular book copy.

The network is a combination of these partial perspectives (i.e., from individual actors, such as reader, library, stock manager, publisher) and a cumulative of chains representing mutual interest (each actor is part of at least a particular chain).

A possible constructional model for the Library in DEMO is presented in Fig. 4. Note the differences in terminology that may result from a different narrative, reflecting the partial perspective of a specific actor. For instance, the economic actor Book Reader from the value model is modelled as CA02 (aspirant member) and CA04 (Member), distinguishing actor roles belonging to different states in the lifecycle of the subject that implements those actor roles.

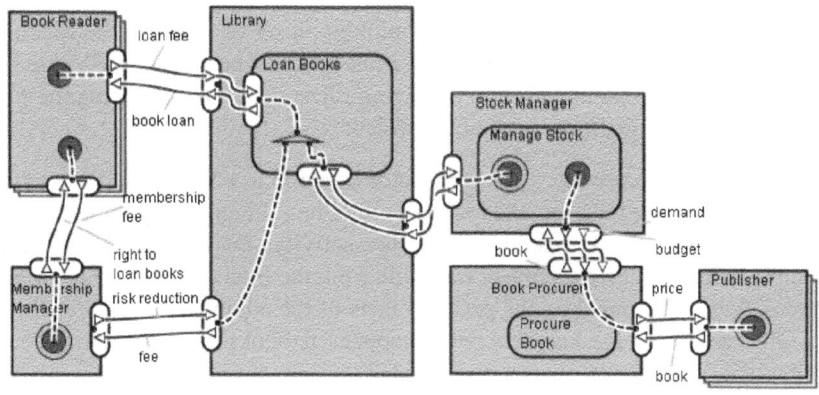

**Fig. 3.** Library e3Value Model

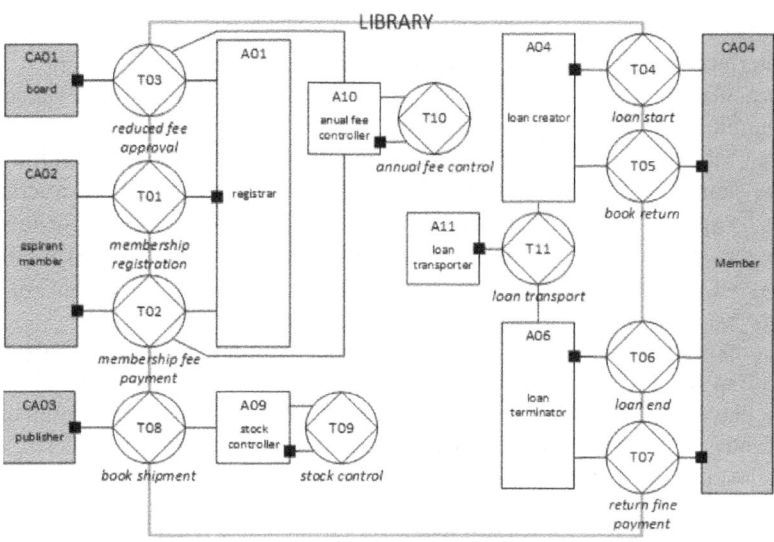

**Fig. 4.** Library DEMO Model

**Value Object.** A value object is defined by Gordijn as "a service, a product, or even an experience which is of economic value for at least one of the actors involved".

The first step is to define which value objects are to be exchanged and, more importantly, why. We find an important relation between *production fact* and *value object* classes since the production of each transaction determines its effective contribution to the value chains it participates in. Hence, a production fact must be about a certain value object, otherwise it is unjustified. The converse is also true: a value object that is not supported by a production fact effectively does not exist.

**Example:** investigating what is the value brought about by the *stock controller* yields two different contributions. The first contribution is title availability management and consists in combining title demand information with the current stock (considering the non-returned books) to minimize the number of *declines* of the *loan book* transaction due to unavailability of the title. The second contribution is actually ordering the titles that were identified as being on-demand efficiently. To model these contributions, two new value objects were identified: *book demand* and *budget* (cf. Fig. 3). The integration of these two value objects in the model implies the identification of additional constructs, namely actors and transactions, as presented next.

**Actor.** Actors are the active elements of both social systems and value networks. In DEMO, an actor is a subject fulfilling an actor role in a transaction type. The initiator and executor actor roles of a given transaction are bound by their common interest in bringing about a production result. In e3Value, both actors (producer and consumer) are bound by the willingness to share value objects with the concept of reciprocity. Note that we will use the *value actor* terminology in the matching ontology for consistency. The specification of value actors is structural in the sense they belong to the network because they are producers or consumers of value objects. Therefore, their value interface (or competences, as described next) characterizes the actor.

A particular subject instantiating the actor role sets the transactional valuations of the economic actor (also including investment and expenses). Then, in addition to the structure of the actor role, it is the instantiation with a subject that allows fully specifying the economic actor. For instance, the Publisher can be specified structurally as an actor by specifying its interfaces (willing to exchange book copies for money). It is, however, only when the attributes of the economic actor are set (that is, when the valuations of its interfaces and the costs of the actor itself are defined) that we may be considering a subject (a particular Publisher, in this case).

This important distinction allows defining the economic actor from e3Value as a composite of both actor role and subject concepts of DEMO. It is in the possible differences between these two that may lay the gap between design and implementation. After identified through consistency verification, this gap may be corrected with alternative solutions, be it different actors or even solutions (expressed as value chains).

**Example:** The matching ontology forces the definition of the economic actor in the value model that corresponds to the *stock controller* actor (A09) in the ontological model. Besides the actor role that is specified by the ontological transaction it takes part in (*manage stock*), A09 has a value interface (defined in the next section) and performs a value activity, let us name it *procure book*. Both are now presented and made clear as separate actor roles in Fig. 3.

**Value Transaction.** In DEMO, a transaction is a sequence of acts following a specific pattern between two actors. One of the actors assumes the role of *initiator* and the other assumes the role of *executor*. For instance, in requesting a book, the Book Reader is the initiator and the Library is the executor.

A *value interface* belongs to a given actor and is composed by a set of value ports (at least one in each direction). The concept of connecting value ports is found in e3value as a *value exchange*. However, is not an exchange in fact, but only a flow (or transmission) of exactly one value object. In this concern, we propose the concept of *value transmission* as a better notion to express the flow of a value object from one actor to another. In a *value transmission*, the initiator has an inbound value port and the executor has an outbound value port.

It is a minimum of two value transmissions in each direction of a value interface that will constitute a *value transaction* and, thus, a value exchange. This specification more clearly enforces the notion of *economic reciprocity*. As an example, the Book Reader has an interface with a value out port for *loan fee* (money) and a value in port for *book loan*, meaning he is willing to make that exchange.

Note that nothing in DEMO forces the existence of a payment or any other kind of reciprocal action by the initiator. As such, the essential model does not warrant completeness from an economic viability standpoint. We will deal with this aspect in the following section, by establishing the need to address it via action rules.

As an additional note, a DEMO transaction should not be mistaken with a value activity. A value activity is encapsulated inside the actor. It is the actor interface that has social meaning; the value activity specifies the origin of the value object but does not have a direct role on the transaction between actors and the related acts.

**Example:** In order to model exactly how the *stock manager* participates in the value chain, its value interfaces and value activities must be specified. In this case, action rules would specify the operational dependencies and order of execution, for instance between making *budget* available to the buyer and effectively order the *book*.

## 4 Value-Oriented System Development Framework Overview

In this section we describe the fundamental components of our approach, beginning with two conceptual principles we elected to follow in devising it: essential value modelling and creating the network from a set of partial perspectives. Next, we present the ontology extension with the core concepts, namely Value Network and Result Tree, and support concepts that materialize the integration with the remainder concepts in the value and construction models. Value Equations are then presented as a base for designing and managing viability of the network, and the section closes with a brief analysis of the role such an integrated ontology has on the value-oriented system development process to support rational organizational decision modelling.

### 4.1 Value First: Value Activities and Value Objects

As explained in section Problem at Hand, one of our objectives is defining an Essential Value Model, which models what is exchanged and why, and not how or by whom it is exchanged. When using DEMO, we are discussing modelling the ontological organization, following the B-I-D distinction. In this case, we need to take a step further, abstracting from the actors and focusing on the structure of production facts and their relations. Indeed, in our approach, designing the chains of production facts takes precedence over specifying the borders of the organization, in order to design those chains in an intellectually manageable and useful way. For that purpose we use a Result Tree model, which has the characteristic of being formulated from a problem/solution perspective, as we are about to describe.

After obtaining such an Essential Value Model, composed by Value Activities and Value Objects, Value Actors are defined by their value object creation and transformation capability, which is commonly referred to as the value he or she adds. At this stage, trivially, each value activity is performed by an actor, in a one-to-one correspondence. Next, a set of value activities from different result trees can be joined in a single value actor by two reasons: 1) they are found to be the same activity (possibly with different names in each result tree) and are joined on value actor that corresponds to an elementary actor role or 2) they are found to be synergetic and are joined on a value actor that corresponds to a composite actor role. The assignment of Value Actors is then completed by their instantiation and valuation of interfaces, which closes the necessary conditions for Value Network model instantiation and deriving viability equations and consequent evaluations for every participant in the network.

## 4.2 Whole Emerging from the Parts – Relativity and Partial Perspectives

Our approach has as a main tenet to deliberately abstract from organizational frontiers at early design time, instead focusing on value activities and value objects. This feature allows modelling both internal and external networks using the same concepts, which provide additional degrees of freedom regarding chain positioning and insight over the modeling of internal actors.

Each participant in a value chain has a specific perspective over the chain. For instance, a for-profit publisher looks at a chain as a solution to obtain revenue in exchange for published books. As he does not possess all the necessary skills to produce published books by himself, he needs other actors to collaborate in producing the published books. As a consequence of economic reciprocity and value model completeness, each node is always both a using system and an object system (in GSDP terminology). This fact implies that it is possible to create actor-specific perspectives using the value model as a base. Such transformation begins by identifying the start stimulus of a given actor and the chain that forms to meet them. The corresponding result tree is then modelled, representing the perspective of the demand generated by that particular actor. The perspectives are then combined by identifying shared actor roles and (possibly) adjusting the Value Object and Value Activity specification accordingly, as part of integrating different Result Trees into a Value Network.

## 4.3 Value Network Ontology Specification

The structure of a value network is compatible with the formal system definition from Enterprise Ontology [5], which defines the following properties: *composition* – a set of elements of some category; *environment* – a set of elements of the same category, disjoint from the composition; *production* – things produced by elements in the composition and delivered to the environment; and *structure* – a set of influence bonds among the elements in the composition, and between them and the elements in the environment. For the production of the matching ontology, presented in Figure 5, we chose to use the World Ontology Specification Language [17], a derivative of the Object Role Modelling (ORM) language [18]. Due to the inherent preciseness and first order logic predicate behind ORM and the expressive power of the predicates connecting classes, this language – and, consequently, WOSL – was found to be an appropriate choice for our goal of maximum expressiveness and minimal ambiguity.

A central conceptual structure while designing a product offering is that of a tree of solutions or results that will converge, in the root, as a result or solution for a certain problem the consumer has, thus our class RESULT TREE (RT). Namely he or she requires a certain object – class VALUE OBJECT (VO) – for his or her activity. In turn, the object provided will be generated by a certain activity – class VALUE ACTIVITY (VA) – which in turn will require one or more objects and so on. Thus, in our ontology we specify classes REQUIRE LINK (RL) and GENERATE LINK (GL). These are existentially dependent on classes RT, VA and VO. We could have modeled these link concepts as fact types connecting directly classes VA and VO. However, it was necessary to create this indirection or apparent redundancy, for reasons that will be clearer ahead.

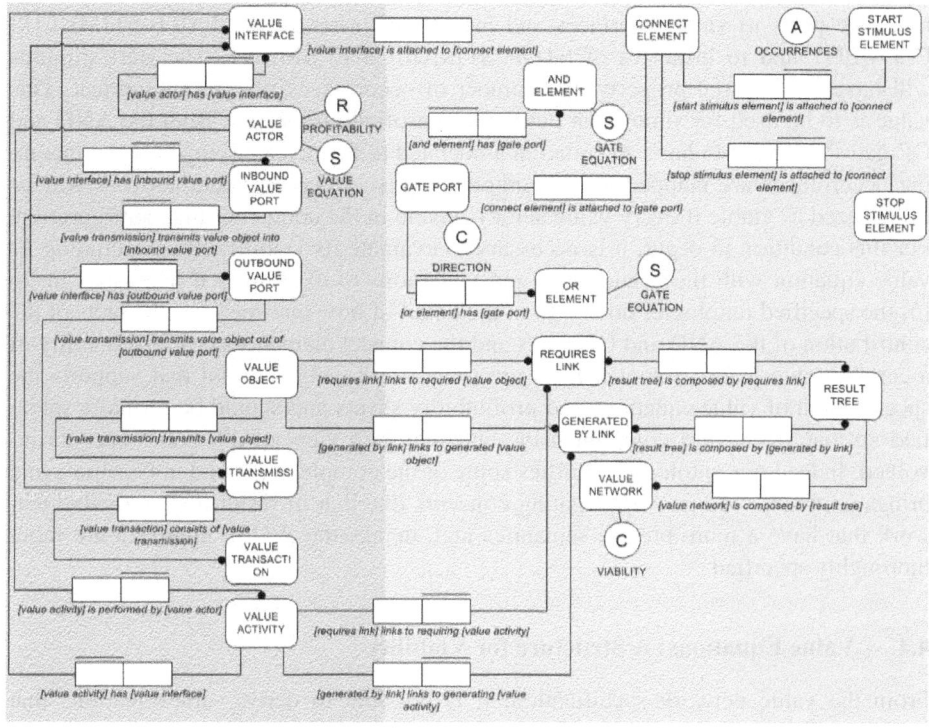

**Fig. 5.** Value Network Ontology – ORM model

Instantiating our ontology in the library scenario, in the context of a result tree (RT01), let's consider that a person may have a VA01 of reading drama material. VO01, temporary possession book copy is a possible object this VA will require. For that we create RL01 that links VA01 and VO01 being part of RT01. VO01 is generated by VA02 book loan. For that we can have GL01 linking VO01 and VA02 also being part of RT01. VA02, in turn will require VO02 book in stock and VO03 risk reduction. Thus the instances RL02 and RL03 linking, respectively: VA02 to VO02 and RT01; and VA02 to VO03 and RT01.

A set of result trees may compose a viable VALUE NETWORK (VN). A VN is composed of one or more trees; if it is composed by more than one tree at least one value actor (performing a VA) has to be shared between each tree and the rest of the network. The need for having classes RL and GL becomes clear now. They allow us to reuse a certain VA – VO pair in different trees. For example the *risk reduction* VO could be reused in the result tree for loaning specially crafted and branded library chairs.

To have a complete picture of the ramifications of the result trees, we need to thoroughly specify how value interfaces connect between themselves with instances of AND ELEMENT and OR ELEMENT. Both of them will have at least 3 instances of GATE PORT being the case that one of them will have outbound DIRECTION and the others will have inbound DIRECTION. Instances of CONNECT ELEMENT will link gate ports to value interfaces and also to an instance of START STIMULUS ELEMENT and to instances of STOP STIMULUS ELEMENT. The start stimulus will have an essential property: the number of occurrences of the value object. This value is to be used for simulation purposes in profitability sheets. Both the AND and OR gates will need to have an equation associated to them, specifying how the incoming occurrences are mapped to the outbound ports. A certain value network will be categorized as viable if every value actor is viable in the context of that configuration. For this condition to occur, it is necessary to evaluate its viability by instantiating its value equation with the occurrences and valuations of exchanges it takes part in. In [6] the specified ontologies do not go to the detail of how one specifies the facts of the contribution of the AND and OR gates and the connect elements to the profitability of a certain value actor. But such facts are found in the e3value tool that supports the specification of value equations and profitability sheets and should be formally specified so that we have a comprehensive and thorough view of all main concepts involved. Indeed our ontology simplifies some of the complexity found in Gordjin's and brings on details and new aggregating concepts like that of result tree and value network that have a more precise semantics and, thanks the WOSL approach are more thoroughly specified.

### 4.4 Value Equations: A Structure for Viability

From the value network specification, it is possible to derive value equations that must be fulfilled by the implementation of the system.

Each value actor has a characteristic value interface specification in the form of value equation. This equation is composed of positive and negative value arguments. As positive arguments, we find inbound value ports. As negative arguments, we find outbound value ports. For instance, value actor Book Reader has the following value equation:

$$V\ book\ reader = book\ loan - book\ loan\ fee + right\ to\ loan\ books - membership\ fee$$

For a full network optimization in our example, we would need to solve the following set of equations to maximize profit of all the actors:

$$\begin{cases} V\ book\ reader = book\ loan - book\ loan\ fee + right\ to\ loan\ books - membership\ fee \\ V\ membership\ manager = risk\ reduction\ fee - insurance\ (risk\ reduction/transference) + membership\ fee - right\ to\ loan\ books \\ V\ library = book\ loan\ fee - book\ copy\ loan + risk\ reduction - risk\ reduction\ fee - book\ stocking\ fee + book\ copy\ loan \\ v\ stock\ manager = book - demand - budget - book + book\ stocking\ fee \\ v\ book\ procurer = book - book\ price - book + demand + budget \\ v\ publisher = book\ price - book - publishing\ costs \end{cases}$$

It is important to note that the value model is valid for a given time period. Different value models can be specified for a series of consecutive time periods.

Having defined the variables and the structure that defines their relations, we say that the value network is viable if the valuations of the value objects that make up the interfaces of each value actor in the configuration make them collectively viable according to the value equations. To instantiate these equations, it is necessary that each of these parameters is multiplied by the number of occurrence defined in the *stimuli*, at a valuation defined in each value port and considering the split percentage in the gates, should they exist.

Due to space constraints, it is out of the scope of this paper to define and exemplify the sub-organization that creates and weaves the value network. Briefly, it consists in applying the 8 rules presented in section 3 and the detailed ontologies to each pair of nodes in the network. This application takes place in the context of a sequential process that creates coherent result tree, value, service and transactional models of each node and the neighboring ones, operating recursively and leaving a structured and robust log of the system change decisions and their rationale in terms of value. Full integration of the complete ontology with the Value-oriented Solution Development Process (VoSDP [7]) is currently ongoing and will be presented as future work.

## 5 Conclusion

In this paper we presented a contribution to Business Modelling and Enterprise Engineering integration state of the art. The alignment between purpose and construction may appear straightforward in complete, static models, when previous knowledge about the business model and its operation are available. However, the fact is that by inspecting any ontological model by itself, we find that all ontological transactions take place at the same level, in an essentially flat structure. This severely hinders evaluating scenarios since it is impossible to discern any depth in terms of intertwined value chains and what were the value conditions that were taken into account in assembling such configuration.

To address these issues, our proposal includes:

- Formal specification of the integration ontology;
- Detailed matching in order to take advantage of specific semantics from both sides, such as competences and action rules specification;
- Formalization of e3Value concepts, such as start stimulus, end stimulus, gates and scenario paths in an ontology for system development;
- A basic set of rules that can be used to check model quality and coherence;
- Bidirectional alignment between purpose and construction, as it is both possible to begin with a value network model and check if a construction model is consistent and the other way around.

Our contribution extends existing e3Value concepts by taking a different perspective directed at supporting a system development process. Our approach consists in isolating and focusing on the value activities and value objects that are successively linked in response to a need of producing a certain result. Providing construction detail in transactional specification may, in turn, prompt for new system development cycles while logging the respective value rationale in the process. It is, therefore, more adequate for using in a system development process that handling change scenarios and reuse of solutions based on a certain configuration.

Furthermore, we specified Value Network as a composition of compatible and viable Result Trees. Value network viability is now rigorously defined as a derived fact from the partial viability of its composition, that is, a certain value network configuration is viable if every value actor is viable in the context of that configuration.

The possible applications of these results to practice are varied. Namely, the structure and methodology Benefits Management [19] can be improved if we specify the relations between the systems out of which business and ICT initiatives arise. The initiatives are specified as addressing gaps between as-is and to-be value and construction models, such as creating a more effective ordering system, or using the existing construction to address new customer segments. Particularly, the matching ontology has been used in practice as the base of a method for supporting Benefits Management [19], as reported in [20]. These results contribute to Enterprise Architecture state of the art since we can formalize the business layer and provide an objective specification of motivations for lower layer components, such as business services and applications, in a rational and traceable way.

Additionally, we contributed to improve DEMO modeling of system networks by providing the concepts that allow adequate discrimination of the system components of individual chains. Such discrimination allows us to specify system composition rationale and (more) objectively define the so-called functional perspective that supports the teleological dimension of a system.

With our contribution, we aim at a paradigm shift in DEMO based enterprise engineering efforts. These efforts have been of a certain "purist" stance in the sense that we can have pure "objective" and "essential" models of reality. However, one cannot talk about purely ontological and purely objective observations of reality. Humans are always constrained by available resources and the goals they have in mind. So models produced will always be biased by a certain motive and dependent on implementation issues, including the resources available for the modeling process itself. Instead of ignoring these so-called subjective concepts, it is time to bring them into play by clearly specifying their relations with the so-called objective concepts. The resulting value-oriented models will then be suitable to address change by supporting discussion, design and implementation of new configurations of value networks.

## References

1. Vargo, S.L., Lusch, R.F., Akaka, M.A.: Advancing Service Science with Service-Dominant Logic: Clarifications and Conceptual Development. In: Handbook of Service Science. Springer (2010)
2. Laudon, K.C., Laudon, J.P.: Management Information Systems: Managing the Digital Firm. Prentice-Hall (2011)
3. Kaplan, R.S., Norton, D.P.: Strategy Maps: Converting Intangible Assets Into Tangible Outcomes. Harvard Business School Press, Boston (2004)
4. Henderson, J.C., Venkatraman, N.: Strategic alignment: leveraging information technology for transforming organizations 32(1), 4–16 (1993)
5. Dietz, J.L.G.: Enterprise Ontology: Theory and Methodology. Springer (2006)
6. Gordijn, J.: Value-based requirements Engineering: Exploring innovatie e-commerce ideas. Vrije Universiteit Amsterdam, Amsterdam (2002)
7. Pombinho, J., Aveiro, D., Tribolet, J.: Value-oriented Solution Development Process: uncovering the rationale behind organization components. In: Proper, H.A., Aveiro, D., Gaaloul, K. (eds.) EEWC 2013. LNBIP, vol. 146, pp. 1–16. Springer, Heidelberg (2013)
8. Simon, H.: The Sciences of the Artificial, 3rd edn. MIT Press, Cambridge (1996)

9. Skyttner, L.: General Systems Theory: Problems, Perspectives, Practice, 2nd edn. World Scientific Publishing Co. Pte. Ltd., Singapore (2005)
10. Dietz, J.L.G.: Architecture - Building strategy into design. Netherlands Architecture Forum, Academic Service - SDU, The Hague (2008)
11. Kundisch, D., John, T., Honnacker, J., Meier, C.: Approaches for Business Model Representation: An Overview. In: Multikonferenz Wirtschaftsinformatik 2012 (2012)
12. Osterwalder, A.: The Business Model Ontology - a proposition in a design science approach. Universite de Lausanne (2004)
13. The Open Group, Archimate 2.0 Specification. Van Haren Publishing (2012)
14. Meertens, L.O., Iacob, M.E., Jonkers, H., Quartel, D.: Mapping the Business Model Canvas to ArchiMate. In: 27th Annual ACM Symposium on Applied Computing, Riva del Garda (Trento), Italy (2012)
15. Kinderen, S., Gaaloul, K., Proper, H.: Bridging value modelling to ArchiMate via transaction modelling. Software & Systems Modeling, 1–15
16. Pombinho, J., Aveiro, D., Tribolet, J.: Aligning e3Value and DEMO – Combining Business Modelling and Enterprise Engineering. In: 8th International Workshop on Value Modeling and Business Ontology, Berlin, Germany (2014)
17. Dietz, J.L.G.: A World Ontology Specification Language. In: Meersman, R., Tari, Z., Herrero, P. (eds.) OTM-WS 2005. LNCS, vol. 3762, pp. 688–699. Springer, Heidelberg (2005)
18. Halpin, T.: Object-Role Modeling: an overview (1998), http://www.orm.net/pdf/ORMwhitePaper.pdf (cited)
19. Ward, J., Daniel, E.: Benefits Management: How to Increase the Business Value of Your IT Projects. Wiley (2012)
20. Pombinho, J., Aveiro, D., Tribolet, J.: The role of value-oriented IT demand management on business/IT alignment: The case of ZON multimedia. In: Harmsen, F., Proper, H.A. (eds.) PRET 2013. LNBIP, vol. 151, pp. 46–60. Springer, Heidelberg (2013)

# ECO-FOOTPRINT: An Innovation in Enterprise System Customization Processing

Yun Wan[1] and Vishnupriya Kalidindi[2]

[1] University of Houston - Victoria, Sugar Land
Texas, United States
wany@uhv.edu
[2] e-CO Matrix LLC., Missouri City
Texas, United States
kvpriyas@gmail.com

**Abstract.** In the overall ownership cost of enterprise system, the maintenance cost consists of a major percentage. During the lifetime of an enterprise system, process customization is the most frequent maintenance efforts. However, current processing method has limited scalability and efficiency. In this case study, we explained how a scalable and efficient customization processing method was implemented. This method used the carbon emission trading mechanism to facilitate the cost benefit analysis of customization request. It also used distributed processing principle to improve the overall processing efficiency. Feedback from a pilot implementation in a large manufacturer included.

**Keywords:** ERP customization, cost benefit analysis, distributed processing.

## 1 Introduction

Enterprises in the United States have undergone two major management changes since 1990s. One was the reengineering of business process, started from 1990s, with the gradual proliferation of distributed information technology, especially PCs and local area networks, into enterprise. The other, started from 2000s, was the so-called eBusiness and the integration of business process with the Web, which is still going on. Underpinning these two major paradigm changes were the frequent upgrading and customization of existing enterprise systems to accommodate changing business environments and competition needs. Thus, enterprise system customization became a major cost factor for system maintenance in recent decade. It is a constant challenge for internal IT to cope with budget constraint and user demands.

Though enterprise system customization is an important IT service, there is relatively little improvement in the processing of request, which is essentially a centralized first-come-first-serve (FCFS) method. In a typical scenario, requests from users are reviewed and aggregated to the IT department. System analyst and IT managers serve as the liaison between business users and developers to translate the request into specific logic design. Then the requested solution is developed and implemented.

Such practice is effective when there is adequate IT budget and moderate number of requests to be processed. However, in the last 20 years, there was increasing adoption of eBusiness practices in enterprise system by large companies, especially the integration of ERP, SCM and CRM systems. E-Business adoption intensified competition and pressure on post-implementation customization, which led to significant maintenance cost of enterprise systems [1]. Thus, an ERP team in large enterprise is often overwhelmed by customization requests to accommodate the constantly changing business environment [2, 3]. When there were too many requests in the waiting line, office politics and influence from executive board are frequently employed to prioritize important requests, which further aggravate the situation.

So a combination of centralized practice, budget constraint, and competition pressure makes the post-implementation, essentially the maintenance and customization of existing system, becomes increasingly challenging [4, 5]. In this case study, we explored this important IT managerial issue through a field investigation on the customization processing practice by a large energy equipment manufacturer in Texas. We explored the limitations of current practice, analyzed the core conflict and challenges, and then presented a distributed new practice inspired by emission control, as well as its pilot implementation in this company.

## 2 Problem Analysis

### 2.1 Company Background

Houston-based 60 year old XYZ Inc. is a public-traded manufacturing company specializing in engineered-to-order equipment and systems in electronic power products and process control systems. Until 2006, the company was using a variety of information systems at its various divisions, a typical information archipelago status experienced by many traditional manufacturers [6]. To integrate the IT infrastructure, the company began to explore eBusiness solution in 2006. By 2011, the company's ERP team was able to successfully implement an Oracle-based ERP platform across the company.

Though the implementation process is relatively smooth and the ERP team was able to work with the various business divisions to re-engineer obsolete practices and adopt the process offered with the ERP package [7], in many scenarios, the team had to customize standard ERP processes to better serve the company's unique needs because the company's manufacturing processes have many special requirements. After the implementation, customization became a major maintenance routine. Every year, there were around eight hundred customization requests being submitted and six hundred being implemented. With the continuous business growth of the company and constantly changing business processes, the team expected there would be increasing number of customization requests generated in the next few years. To better serve this increasing need, the IT department was actively looking for ways to improve its current customization processing practice.

## 2.2 The Existing Practice

To better understand the current customization processing practice within the company, authors of this paper conducted a field study and started by classifying current incoming requests.

There were three categories of ERP customizations requests being identified in our field study. The first category deals with those customization needs that can be easily solved by configurations. To accommodate the diversified needs, most ERP software packages come with configuration options without any invasive modification to the source code. The second category of customization involves configurable workflows and metadata drive personalization, which are enabled in most ERP software. With the use of these methods, a business process in ERP can be tailored to a company's needs without custom code. The third category of customization, also the most complex one, involves the use of extensions to accommodate special needs of the company, such as interfacing with other information systems, revenue recognition, check printing, custom purchase order layouts, etc. This category of customizations involves developing reports, interfaces, conversions, and extensions onto an existing ERP system. The third category customizations sometimes involve invasive modifications to the ERP design or source code to change how it functions so as to fit into the company's business process. This is not supported by the vendor and is a risk in terms of future upgrades. Though customization under third category are complex to build and can sometimes be risky, about 40% requests in the company came under this category because they could potentially bring in more competitive advantage on business processes for the company.

Because customization is a necessary evil, the ERP team has established a centralized processing practice to evaluate whether a customization has marginal benefit to the company and if it could be completed on time and within budget as illustrated in Figure 1.

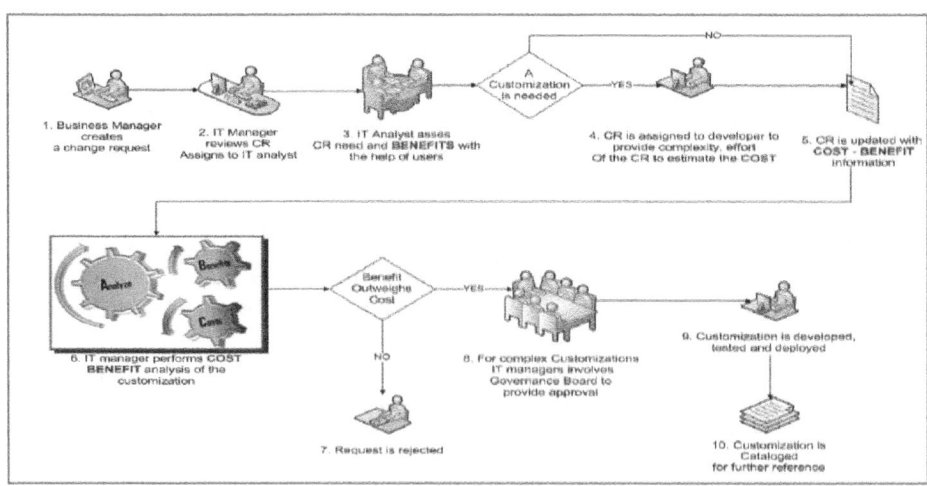

**Fig. 1.** Current customization practice

In current practice, IT managers conduct an initial review of all incoming requests and then forward the request to IT analyst and ERP team to work with business users on the necessity of the customization, including the assessment of the cost and benefit. Once the analysis has been done, IT managers make a decision on whether to implement it or not. For high complexity or risky request, IT governance board would be involved for final decision. This is a common practice adopted by many large companies.

## 2.3 The Challenges

Through investigation, we found though there were merits with current practice, the major limitation was its lacking of *scalability*. The current practice cannot accommodate the increasing number of requests to be processed, which led to the company's various business divisions experience delays in request processing. As a result, the ERP team was under constant pressure and their contribution was being discounted because of user dissatisfaction.

We also found that despite the frustration of both users and ERP team, from time to time, some unnecessary and costly customizations were eventually implemented after waiting in the pipeline long enough, then had to be reverted after a while. Similar phenomenon was also observed in other studies [8]. This indicated the current practice could become ineffective when under pressure. That is, once being overloaded with requests, the current practice may no longer be able to sort out unwanted requests effectively. Upon further analysis, we identified three conflicts led to the above scalability limitation.

Firstly, while there were limited IT resources to expend, there were no internal or external restrictions and/or incentives on company's business users' submission of new request. This conflict was mainly due to the fact that the ERP team served business users' customization needs without incurring cost-related accounting, so though business users have to spend time with IT analysts working on customization request, they have little incentive to restrict their customization requests, no matter whether the request could increase margin value to the company or just for their convenience.

A second related conflict was the limited IT resources and the lack of measure to make business users objectively prioritize their requests. All business users wanted their request to be processed quickly. They were also aware of the limited overall IT resources that can be applied towards the customization requests at any moment. Thus, they tended to emphasize the importance of their requests with more or less exaggeration to push their request through. With inflated priority signals from many business users, the ERP team had difficulty to give priority on important request, thus FCFS method has to be used to treat all incoming requests equally, though this was less efficient. In addition, current system provided little flexibility or convenience to allow business users dynamically adjust the priority of their request in the pipeline.

A third conflict came from the information asymmetry between business users and IT staff in terms of evaluating customization cost and benefits to determine return on investment (ROI) for each request. Currently, the benefit of customization was estimated by IT analyst with the help of business users. Similarly to request priority assessment, business users tended to exaggerate the benefits of their requested customization to increase the chance of getting approval. Though IT analysts knew such tendency, they had difficulty to detect its extent because of domain knowledge limitation. Meanwhile, though IT analyst could accurately predict the cost estimation for each request, such information was not shared with the business users, nor did business users have to be concerned about it. Actually, disclosing cost information under current practice would only increase the business users' tendency to inflate the benefit of their request.

Because of above conflicts, the current centralized processing practice has major limitations in scalability in term of handling increasing number of customization requests. Many requests ended up clogging up the overall process pipeline without being processed timely. For truly important request, business users have to leveraging corporate politics and use pressure from governance board to get it moving. This added additional chaos, confusion, and unfairness feeling on many stakeholders.

To overcome the limitation, we need to design a new method to address these conflicts. Next, we explain how we use an innovative new practice to solve this problem.

## 3 The New Practice

### 3.1 Principle of Decentralization

To increase the scalability of a working system, we usually employ market mechanism and distributed method. A fully informed market that allows individuals to make decision based on their self-interests would achieve efficiency automatically [9]. A distributed method would break down complex task and allow individual participants to self-allocate IT resources among themselves with mutual adjustment, thus allowing a system to grow in proportion with the system size without losing order; similar concepts has been utilized in innovative management practice via swarm intelligence and crowdsourcing [10, 11].

In this case, the goal of the new practice was to give users incentive to be self-constrain in request submission, benefit estimation, priority claim, and finally allow the IT department to process the request within budget and on time. Thus we needed to encourage users to reveal the actual estimation of benefit for their request. We also needed to automate the cost estimation for each request with the help of ERP team. With the combination of these two pieces of information, a ROI for cost/benefit analysis would be available for decision-making.

To solve the conflict between limited IT resources and unrestricted request submission, we needed to have the overall requests under control. And within this overall limit, business users could submit their request with and have those truly useful requests with high priority.

The combination of all these goals indicated a "cap-and-trade" model exemplified by carbon emission trading [12].

There were at least two basic emission control trading models. In a traditional "cap and trade" model, an aggregated cap on all emission sources was established, which translate into a specific amount of emission permit. Each polluter or emission source would receive certain amount of emission permits. They were allowed to trade emission permits among themselves to meet their actual emission needs. An alternative approach was a "baseline and credit" model without a cap. In this model, an emission baseline was established. For polluters that could reduce their emissions below their baseline level, they can create permits or credits, usually called "offsets," for other polluters to purchase [13].

The cap and trade model gives government better control on overall emission level. However, it could have adverse effects if the cap amount was set incorrectly or the allocation of permits among participants was not efficient due to lacking sufficient data. In contrast, the baseline and credit model gave government less controlling power but gives participants more flexibility.

When applying the emission trading model to our current practice, we found two connections between the emission control and request processing: firstly, the company has an overall budget cap on system customization in each budget period, which is similar to emission cap; secondly, we wanted business users to be self-constraint in their customization request and submit only useful requests, which is similar to expectation on polluters to only release truly necessary carbons into atmosphere. Since there was a clear budget cap and historical customization requests records from each business division, we chose the cap-and-trade model and permit exchange market in carbon emission trading to be the model and starting point of our new customization processing framework design.

## 3.2  Cost Drivers and Estimation

To create a distributed trading mechanism, we need a cost measurement for customization request. This measurement doesn't necessarily reflect the actual cost of a customization. However, it has to reflect the relative cost differences among requests. The major cost drivers for an ERP customization includes task size, complexity, and risk level.

Traditional cost estimation methods use line of code (LOC) and function points (FP) to measure task size. Later, with proliferation of objective oriented programing and component-based software development method, software component like objects, modules and use cases were also used to estimate cost [14]. In our case, the customization requests come in many different formats, including code, GUI screen, workflows, report layouts, etc. So it is difficult to use empirical method like COCOMO II [15] or related OO software estimation techniques like those suggested in Lorenz et al. [16].

Complexity is a second cost driver for customization. The complexity cost drive is related but independent from task size [17]. A request with many GUI and report layouts may be far less complex than a small but invasive revision on source code because the latter may have significant impact on other components, the package that contains the revision, and the platform it resides.

Risk level is the third cost driver for customization. Customization risk is related to both task size and complexity [18]. It is also determined by the category of customization and the type of revision involved. In our system, risk is determined by the category it involves and the specific customization type.

The measurement of customization cost has to include all cost drivers and their interactions. So we decided to use a holistic method to measure the cost. Basically, we identified customization category, type, functional track it serves, as well as operating system and platform that are involved. All these factors determined the approximate measurement of cost drivers we identified in a request. Meanwhile, since the company has a rich archive of historical customization cost data with these factors, we used historical data to predict the cost measurement for new request.

Thus, in the new processing framework, we asked business users to identify above key factors when submitting their request and then used decision-tree algorithm to predict the cost of the new request. After obtaining the predicted cost, a unit of measure was created - it derived from the cost drivers behind customization. We named it "e-CO Footprint" to reflect its connection and analogy with emission and its control.

### 3.3 The Exchange

Drawing inspiration from carbon emission trading, an electronic exchange market ("Exchange") was created for the new system. A concept of "customization processing allowance" or "allowance" in short, equivalent to permit in emission control, was established. Certain amount of such allowances is allocated to business users at the beginning of each budget cycle. This allowance matches pre-allocated e-Co Footprint cap assigned to each business division.

In European Union, the allocation of carbon emission allowance was conducted in two levels, EU and state members. The EU was responsible to define overall emission cap and the cap for each member state. Each member state worked with its various participating companies to decide the allocation of allowance among them, usually through the coordination of the environmental as well as economics and trading ministries [13]. In this case study, each user division would be assigned equal amount of allowances. In the future implementation, the initial allowance could be allocated based on historical requests data and adjusted by strategic priority of the company. The latter can be decided by IT management and governance board.

During each budget period (like quarterly IT builds), users in each business division or functional track, such as manufacturing, finance, human resources, project management, service, etc., could submit their customization requests via the Exchange.

Based on customization complexity, processing cost is derived in the unit of e-CO Footprint. As long as a business user still has available allowance to cover the cost, the customization request is processed through regular operational procedures. Once submitted, the customization cost is deducted from the business user's allowance and the request is inserted into the pipeline.

Users in the Exchange are able to observe the status of all requests in the pipeline. This helps the business users to estimate the waiting time for their request as well as the opportunity of borrowing or lending allowance. If a business user is running out of allowance before the budget period and still needs to implement an important request, they could either withdraw request in pipeline that haven't been processed to recover enough allowance, or borrow allowance from a common pool in the Exchange. If a user has or expect surplus allowances by the end of budget period, they could loan them to the common pool or rollover to next budget cycle. The specific trading policies in the Exchange are designed to be aligned with the overall strategic goal of the company [19], and also be integrated with other company management system like balanced score card [20]. The new customization processing model is illustrated in Figure 2.

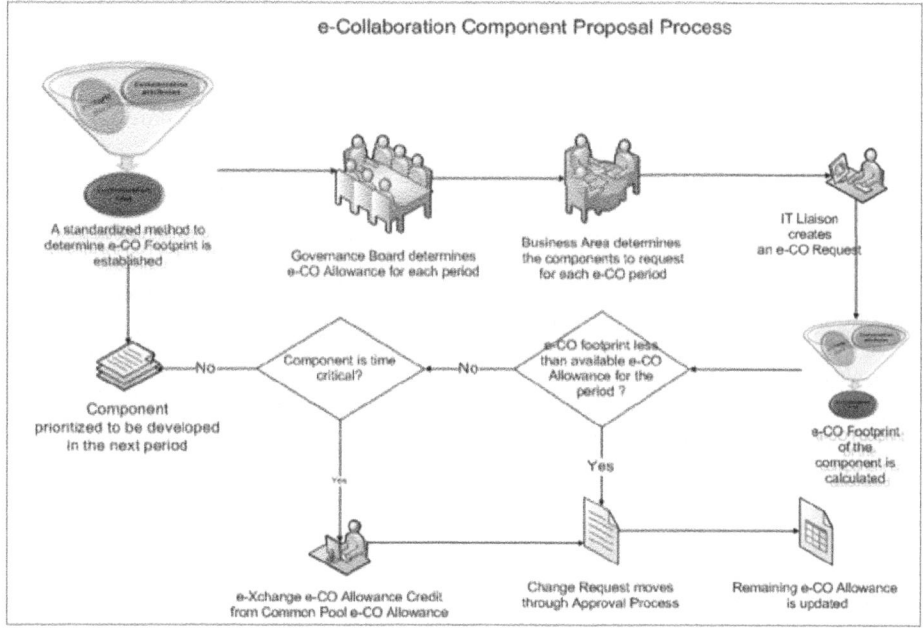

**Fig. 2.** New customization processing model

### 3.4 User Roles and Interfaces

After the framework was created, we developed a prototype of new processing system. To be consistent with our measurement, we named the new system "e-CO Footprint Tracker." It has a flexible design that allows different enterprises to configure the factors affecting their specific customization allowance and cost structure.

The system was developed as a "Software as a Service" on Oracle Cloud to facilitate future revision and coordination. A rapid application development tool called "Oracle Application Express" or APEX was used to develop the system. The database objects, namely, tables, sequences, functions and procedures, etc. reside in the database schema set up in the Oracle Cloud. Oracle's native SQL and PL/SQL scripting language was used to develop the database objects. The screens were designed with APEX development platform.

In addition to a system wide administrator who is responsible for overall system maintenance, users within an organization were assigned different roles according to their position in the customization processing procedures (see table 1). Security features were built into the system based on roles. Once users login the system, they would see a customized dashboard with information relevant to their roles (Figure 3).

**Table 1.** User roles and responsibilities

| Role | Responsibility |
| --- | --- |
| Administrator (system wide) | Overall system administrator, system maintenance and provide support to different organizations |
| Governance Board (organization wide) | Determines the factors and quantifiers related to e-CO Footprint calculation of an organization |
| Super user (organization wide) | Responsible for e-CO Footprint tracker configuration of an organization |
| Requester (organization wide) | Responsible for entering e-CO proposals that help the organization utilize the enterprise system tools efficiently. |
| Reviewer (organization wide) | Responsible for making decisions on e-Co proposals related to certain business area |
| Approver (organization wide) | Responsible for reviewing and approving e-Co proposals from all business areas of the organization. |

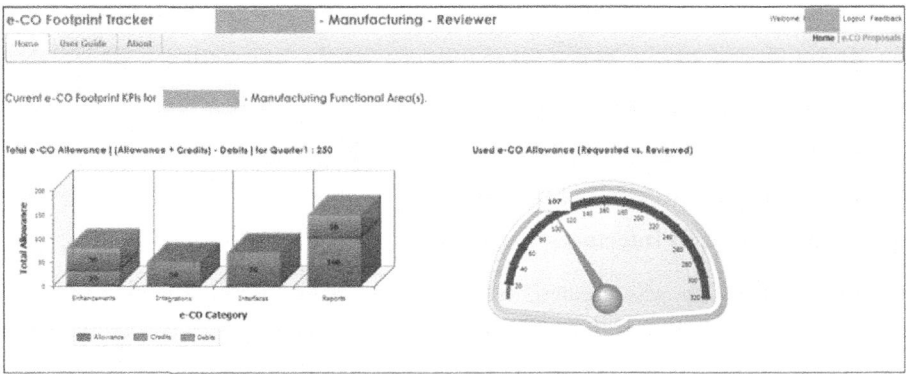

**Fig. 3.** The reviewer dashboards

ECO-FOOTPRINT: An Innovation in Enterprise System Customization Processing    129

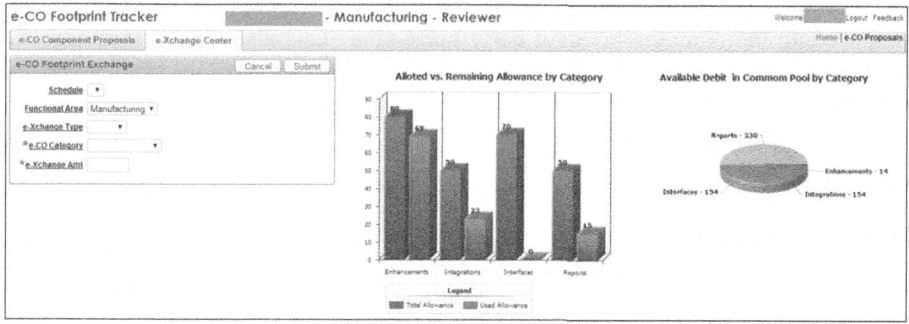

Fig. 3. (*Continued*)

## 4 Pilot Implementation

### 4.1 Implementation Details

With the approval of the IT director, a pilot implementation of the new system was conducted from June 14 to July 3, 2013 in the company. A total of 18 users from five business divisions participated in this pilot implementation.

The super user role was co-served by the IT director and the business process lead for human resource. Both of them are active members of the IT governance board at the company. They were able to make strategic decisions on system configurations, including allocation of allowance among different business divisions for this implementation. The approver role was also served by the IT director because she actively interacts with business users and understands user needs. The reviewer roles were served by five individual employees from different business divisions of the company, including business process leads from purchasing, sales, project management, manufacturing, and human resources. The role of requesters was assigned to a mixed group of company employees from the business team as well as technical and functional staff for the ERP system. They were chosen so they can provide inputs on the process, workflow, as well as the software quality of the system from their perspectives and in different stage of process.

The allowances for each business division were distributed with equal amounts of 250 units in this implementation. This initial equal distribution allowed us to explore if the business users would utilize trading mechanism to lend or borrow these allowance to meet their customization needs. To prevent business users from spending too many allowances in one customization category, the 250 units for each business division were further divided into four customization categories. The distributions were 30 units in enhancements, 50 units in integration, 70 units in interface, and 100 units in reports. This distribution configuration was estimated from historical actual spending of customziation categories. Since a businss user may need to borrow units to file a request during the pilot operational period before other business user would like to lend, we also released 100 units in each customization category into the common pool for early borrowing. This also helped to relieve the risk that we may under-estimaite the total requesting needs during the implementation period.

## 4.2 Data Collection

Since the major purpose of the pilot implementation was to collect feedback on the new method. We designed three ways to collect such information: user survey, system transaction data, and interviews. The system platform would automatically collect the most fine-grained user transaction data. We also designed the system to collect specific transaction data in the database. An online survey was created with ten questions to get feedback on user interface design and the acceptance of new practice. Along with the survey, a few interviews were conducted with users who participated in the pilot implementation and had played key roles in the process. The interviews were recorded on the iPhone with auto transcription.

# 5 Outcomes

## 5.1 Surveys

To assess the design of the system and feedback on the distributed processing mechanism, we conducted a survey after the pilot implementation. There were 9 requesters, 5 reviewers and 1 approver took the survey.

We used TAM theory and its instruments to assess the perceived usefulness and ease of use for the system [21]. The average perceived usefulness for the system was 4.98 out of a possible 6 or 83% and the perceived ease of use was 5.38 or 90%, higher than former. The results were tested for internal consistency, and the Cronbach's Alpha for perceived usefulness was 0.927, an indication of high consistency, which means the system was perceived consistently useful from a user perspective. The same test for perceived ease of use was 0.362, which means there were inconsistent perspectives on how easy this system was to use. This discrepancy could be due to the fact that the users did not receive enough training on the system, and they used it for only a few times. When proper training is provided and the user interface is further improved, it is expected that the ease of use will increase.

In addition, all survey participants felt the system would "likely" make management of customization request processing easier and 86% of them felt it is "extremely" or "quite" likely. 87% of survey participants felt the system has feasibility to be implemented in the company and 53% felt "extremely" or "quite" likely. In contrast, 13% felt neutral about feasibility and none indicated unlikely feasibility, which indicated very positive acceptance to the new method.

In terms of whether this new system could improve efficiency, 87% of them felt it is likely and 73% felt it is "extremely" or "quite" likely, compared with 13% indicate neutral opinion and none indicated unlikely. Finally, 93.3% of them felt the new system brought value by providing visibility in the customization processing and only 6.7% indicated neutrality.

Thus, overall the participants were positive about our new system and the distributed trading concept behind it. Most of them believed this new method can bring efficiency and transparency to the customization management.

## 5.2 Transactions

We also analyzed data from transactions. During the two week operating period, the requesters entered a total of 55 requests and on average 3.9 new requests were generated per working day.

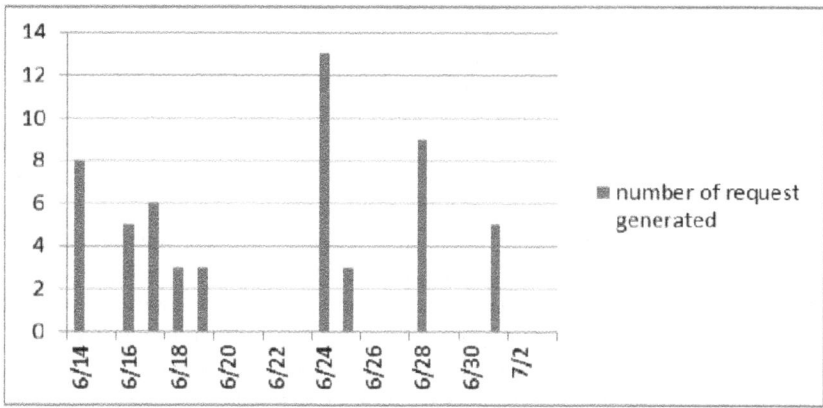

**Fig. 4.** Number of requested generated each day

The reviewers processed 35 of them and recommended 27 requests to the approver. The remaining 4 requests were returned to requester for further revision and 4 others were rejected. Among the 27 recommended requests in the pipeline, 7 transactions were approved and moved into operations by approver and two were rejected hence removed from the pipeline.

As mentioned previously, we initially allocated a total of 250 allowance units to each business division. We wanted to use this simple allocation schema to test the self-allocation efficiency of the trading system. The operation started on June 14. The first lending and borrowing happened on June 24, which was the $11^{th}$ day of operation. The last lending and borrowing activity happened on July 3, the last day before the operation closed. This indicated users' learning curve picked up rapidly and they were actively engaged in trading once being accustomed to the new practice.

During the overall operational period, a total of 677 allowance units had been loaned to the Exchange and 425 units have been borrowed from Exchange. This led to a 1.6 lend/borrow ratio, which indicated we may allocate too many allowance for the current run, a potential waste of resources. Ideally, the ratio should be closer or equal to 1. We have archived specific lend/borrow data from each business division and customization category. The data can be used to optimize future allowance allocation as well as other system maintenance and budgeting purpose.

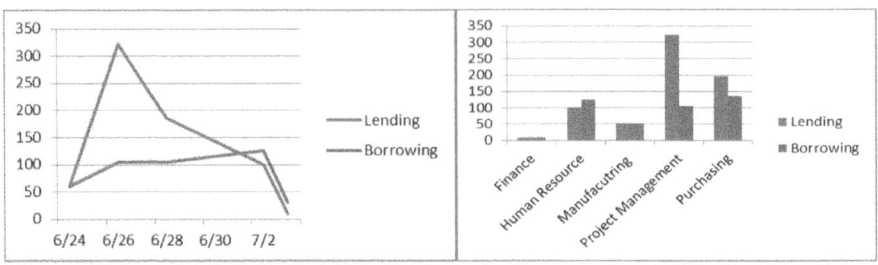

**Fig. 5.** Allowance lending/borrowing activities across dates (left) and business divisions (right)

### 5.3 Interviews

The new system aims at providing a scalable and transparent customization processing method to enterprise. However, we have concerns about whether business users appreciate the new principles and would like to comply with the new system rules. The new rules do put certain restrictions compared with current processing method though both the business users and the company would benefit from the new system in terms of less waiting time and more efficiency.

From survey outcome previously discussed, we received very positive feedback in terms of user acceptance and most participants were requesters. In the interview, we focused on reviewers and approvers because they played most important roles in the new system and were in a better position to compare effect brought by the new method. We found that they not only understand but also appreciate the new method. In addition, they provided many insights we didn't expected.

For example, one reviewer has following optimistic comments on the educational effect of new system on business users. He thought the new practice would force business users to be self-constraint on request submission and proactively estimating the true benefit a request can bring to their business process:

> "People think IT resources are free. So this tool will give us the ability to look at what the different departments are requesting and what type of return on investments will this provide our company. I think this feature will give manager visibly and accountability on things requested by their employees. Did I budget for it? Did I take the time to research the cost impact? If it is that important can I borrow allowance?"

The same optimism was echoed by another reviewer. This reviewer also brought up the importance of providing transparency on overall user activities to business users, so the business user could make informed requests and also make adjustment in their trading behavior:

> "I think the biggest efficiency of this is looking at our resource capacity in one place. We only have so many people and they can only accomplish so many things in a given period of time, so I think at least putting some kind of maximum of what that looks like and quantifying it is helpful and would be really beneficial for a

group, rather than an endless wish list that we can never get to the end of. This process would ensure we are always working on the most important things. Also, the metrics that you have incorporated, all of the reporting to be able to see which groups are going which route and also which groups are maybe giving away their credits more frequently."

We also found evidence that at least a few users had already fully understood the underlying concepts embedded the new system and appreciated the restriction we set on IT resources:

*"Overall I love the concept of the e-CO footprint tracker. I think when any resource is viewed as unlimited be that money, time, effort that the best solutions are not generated. When there are limited resources the most creative & effective solutions will emerge. When there is a limit to what can be done systematically it will keep organization look at a few things. They we'll need to look at the business process as well as the people who are executing the business process. Often times technical solutions are meant to dummy proof and to compensate for a lack of discipline that a person may have. This will force an organization to look at the employees performing process performing transactions to determine are they have the right skills or if they are trained correctly. I think a lot of times there a lot of human resources issues that need to be addressed in order to make something successful before always making it a technology solution."*

From the approver perspective, this new method was a clever combination of carbon emission control principle and software cost benefit analysis. It helps corrected a pre-occupied belief probably held by many IT directors or CIOs that improving customization processing efficiency needs substantive resource inputs.

*"Participating in the study has really opened my mind to a possibility of using a structured approach to streamlining the process of determining which enhancements would better the company. I've always felt this required a significant amount of subjective input that would be difficult to program, but by combining the concept of carbon footprint with a cost-benefit analysis it provides the necessary information to effectively assign the resources to the highest priority efforts."*

The overall response from interviewees confirmed our findings from survey that users have very positive acceptance of this new method. It also indicated that many users grasped the purpose of using decentralized method in the new practice and would like to comply with it. All these outcomes provided strong support and confidence for a formal implementation of the new practice in customization processing in the company.

## 6 Conclusion

Improving efficiency in enterprise system post-implementation maintenance could bring great benefit to enterprise. Enterprise system customization is major maintenance activity and a major cost component. Current centralized customization processing methods have limited scalability. In this case study, we designed and implemented a decentralized method. This method used borrowed carbon emission control mechanism to help business users self-manage their request, self-allocate IT resources, and reduce potential waste in request generation. From a pilot implementation in a large manufacturer for this new method, we found very positive feedback.

The positive evidences support a formal adoption of the system, which is currently being evaluated by the company. The same system and the concept it represents could be easily reconfigured for other company to use in their enterprise system management. In addition, this new method indicated a prospective to be adapted for more diversified IT management needs because of its scalability and cost-saving potential. We are actively exploring this aspect.

## References

1. Ng, C.S.P., Gable, G., Chan, T.: An ERP maintenance model. In: 36th Annual Hawaii International Conference on System Sciences. IEEE, Honolulu (2003)
2. Salmeron, J.L., Lopez, C.: A multicriteria approach for risks assessment in ERP maintenance. Journal of Systems and Software 83, 1941–1953 (2010)
3. Sekatzek, E.P., Krcmar, H.: Measurement of the Standard Proximity of Adapted Standard Business Software. Bus. Inf. Syst. Eng. 1, 234–244 (2009)
4. Law, C.C.H., Chen, C.C., Wu, B.J.P.: Managing the full ERP life-cycle: Considerations of maintenance and support requirements and IT governance practice as integral elements of the formula for successful ERP adoption. Computers in Industry 61, 297–308 (2010)
5. Hecht, S., Wittges, H., Krcmar, H.: IT Capabilities in ERP Maintenance - a Review of the ERP Post-Implementation Literature. In: ECIS 2011 Proceedings (2011)
6. McKenney, J.L., McFarlan, F.W.: The information archipelago–maps and bridges. Harvard Business Review 60, 109–119 (1982)
7. Hammer, M.: Reengineering Work: Don't Automate, Obliterate. Harvard Business Review 68, 104 (1990)
8. Haines, M.N.: Understanding Enterprise System Customization: An Exploration of Implementation Realities and the Key Influence Factors. Information Systems Management 26, 182–198 (2009)
9. Malkiel, B.G., Fama, E.F.: Efficient Capital Markets: A Review of Theory And Empirical Work. The Journal of Finance 25, 383–417 (1970)
10. Bonabeau, E., Meyer, C.: Swarm intelligence: A whole new way to think about business. Harvard Business Review 79, 106–115 (2001)
11. Kittur, A., Smus, B., Khamkar, S., Kraut, R.E.: Crowdforge: Crowdsourcing complex work. In: Proceedings of the 24th Annual ACM Symposium on User Interface Software and Technology, pp. 43–52. ACM (2011)
12. Christiansen, A.C., Wettestad, J.: The EU as a frontrunner on greenhouse gas emissions trading: how did it happen and will the EU succeed? Climate Policy 3, 3–18 (2003)
13. Ellerman, A.D., Buchner, B.K.: The European Union emissions trading scheme: origins, allocation, and early results. Review of Environmental Economics and Policy 1, 66–87 (2007)

14. Pressman, R.S.: Software engineering: a practitioner's approach. McGraw-Hill, New York (2010)
15. Boehm, B.: Software Engineering Economics. Prentice-Hall, New York (1981)
16. Lorenz, M., Kidd, J.: Object-oriented software metrics: a practical guide. Prentice-Hall, Inc. (1994)
17. Alkoffash, M., Bawaneh, M.J., Al Rabea, A.I.: Which Software Cost Estimation Model to Choose in a Particular Project. Journal of Computer Science 4 (2008)
18. Czarnacka-Chrobot, B.: Rational Cost Estimation of Dedicated Software Systems. Journal of Software Engineering and Applications 5, 262–269 (2012)
19. Henderson, J.C., Venkatraman, N.: Strategic alignment: leveraging information technology for transforming organizations. IBM Systems Journal 32, 4–16 (1993)
20. Kaplan, R.S., Norton, D.P.: Using the balanced scorecard as a strategic management system. Harvard Business Review 74, 75–85 (1996)
21. Davis Jr, F.D.: A technology acceptance model for empirically testing new end-user information systems: Theory and results. Massachusetts Institute of Technology (1986)

# Automatic Model Transformation for Enterprise Simulation

Yang Liu and Junichi Iijima

Department of Industrial Engineering and Management,
Tokyo Institute of Technology, Tokyo, Japan
{liu.y.af,iijima.j.aa}@m.titech.ac.jp

**Abstract.** In order to simulate business process models, either an additional mapping schema is developed or the models are translated into other diagrams that can be used as a conceptual model of simulation. However, most existing methods require manual transformation, and they have made troublesome and time consuming when the business process models are complex. Thus, the application of such translation method is limited. In this research, we conducted a Model Driven Framework base transformation to semi-automatically transform DEMO aspect models into DEVS executable model. Contribution of this research could be concluded as: (1) It makes DEMO model "really executable" and becomes more helpful in BPR; (2) It provides a suitable and well supported formalism and semantics for conceptual model for discrete event related enterprise simulation; (3) It helps developer semi-automatically generate a simulation model which provide better support for enterprise simulation development; (4) DEMO based DEVS definition could be implemented in other simulation platform for better practice.

**Keywords:** DEMO, DEVS, Simulation, Model Transformation.

## 1 Introduction

Business process model is always used as a multiple-purpose tool for understanding operations of existing organization to assist business process redesign and reengineering. However, redesign and reengineering always involves changes in people, processes and technology over time. Sometimes, the interactions of people with processes as well as the possible result of changes need to be evaluated and compared. Only modelling may not provide enough information to achieve the objective, this is where simulation can provide value. Several researches ([1][2][3]) have been conducted in business process simulation field. Other researchers ([4][5][6]) also suggested that business process modelling and business process simulation should be combined. On one hand, business process model should be complemented with simulation for significant benefits and results with certain accuracy evaluation. On the other hand, simulation may provide little help without profound conceptual modelling preceding it.

A business process simulation should be based on well-defined formalism and semantics to achieve accurate and precise results. Enterprise ontology[13] describes the operation and construction of an enterprise, which address explanatory and /or predictive relationship in observed phenomena. This ontology has been utilized as a precise conceptual model for enterprise simulation in several researches[4] [7] [16].

In order to simulate business process models, either an additional mapping schema is developed or the models are translated into other diagrams that can be used as conceptual model of simulation. Barjis[4] introduced a concept called "executable model", translating enterprise ontology DEMO into Petri net[7] to make model "executable". Also there are other translations such as from DEMO to XML[8] or DEMO modeling tools like Xemod. However, all these methods require manually transformation among models. The transformation becomes troublesome and time consuming when business process is very complex. Thus the applications of such transformation method are limited.

MDD4MS[10][11], transforming BPMN into DEVS, provides a generic framework for convert concept model to executable simulation model based on model driven framework. This framework is valuable in making business process models "really executable". However, as a conceptual modeling method for enterprise simulation, BPMN lacks of semantic meaning in enterprise level, do not views on collaboration and communication and goes too much into details to simulate in enterprise level.

Based on previous researches, we conducted a model driven framework based transformation to make enterprise ontology, DEMO, executable in a simulation platform, such as DEVS. Outcome of this research including three parts: (1) Meta-models for: DEMO CM, DEMO PM, DEMO AM and DEVS; (2) Modeling platforms for (CM, PM, AM and DEVS) according to the meta-models (3) Transformation rules on meta-models (CM to PM, PM to AM, AM to DEVS S1, DEVS S1 to S2). Contributions of this research can be summarized: (1) it provides a sample on how to conduct enterprise simulation from enterprise ontology, which makes DEMO model "really executable" and becomes more helpful in BPR; (2) it provides a suitable and well supported formalism and semantics conceptual model for DEVS based enterprise simulation. To conduct better practice, This DEMO based on DEVS model could be translated or implemented by the different simulation platform; and (3) it helps developer semi-automatically generate simulation model that can better support simulation development.

The remainder of this paper is organized as follows: Firstly, related concepts o DEVS, enterprise ontology for simulation and MMD4MS framework are reviewed in chapter 2. Next, framework of the research, meta-models, and the transformation rules of are introduced in chapter 3. Finally, a brief discussion and future work are listed in chapter 4.

## 2 Background Theories

### 2.1 Discrete Event Simulation

Discrete event simulation is an effective tool for analyzing and designing complex systems. It is well known as mathematical formalism based on system theoretic principles. Any systems with discrete event behaviors can be represented by the DEVS formalism, and an equivalent DEVS representation can be found by other formalisms.

Classic DEVS specification defines the structure of the basic DEVS formalism. Models expressed in the basic formalism are called Atomic models. The atomic DEVS model is defined by the following information: the set of input values, the set of output values, the set of states, the internal transition function, the external transition function, the output function, and the time advance function. Input and output ports provide an easier way of modeling and an elegant way of building larger models. Coupled DEVS specification defines the means for coupling the atomic DEVS models. The coupled DEVS model defines the following information: the set of input ports and values; the set of output ports and values; the set of sub-models; EIC (external input couplings), EOC (external output couplings); and IC (internal couplings). EIC connects an external input to a component input; EOC connects a component output to an external output; and IC connects a component output to a component input. Hierarchical DEVS extends the version of coupled DEVS that allows coupling both atomic and coupled models [12].

### 2.2 Enterprise Ontology - Conceptual Foundation for Discrete Event Simulation

Enterprise ontology[13] describes the operation and construction of an enterprise, which addresses explanatory and /or predictive relationships in the observed phenomena. Comparing to workflow based business process modeling methods such as UML, IDEF and BPMN, DEMO is more suitable for identifying the conceptual modeling methods of enterprise simulation with the following reasons:

- DEMO describes an enterprise in semantic, therefore, the other modeling methods are in syntactic. Simulation model themselves deal with only specify syntactic concepts. For example, they account for notation of entity, event and state in discrete event simulation; place, token in Petri net, and so on. Semantic definition of real world should be given in conceptual models. Normally, business process models for simulation are designed as tasks or work-flow based such as BPMN or UML. These business process models lead to arbitrary and inconsistent models in conceptual modelling stages of the simulation. On the contrary, DEMO highlights enterprise ontology, dealing with the semantic meanings of the enterprise, as a coherent and consist model.
- DEMO describes not only a workflow in enterprise but also the construction and interaction of social systems. Operations among organizations are complex, collaborative, and interactive phenomena, and there are multiple engaged stakeholders (actors) for communicating, coordinating, and agreeing on certain tasks. It precedes the role of its members, the responsibility and social connections, rather than

approaches in conventional methods. Therefore, SAMPO[14], DEMO[13], BAT[15] and others used a new framework based on LAP. DEMO, an enterprise ontology founded on ψ-theory (PSI: Performance in Social Interaction) of Enterprise Engineering[16], explains how and why people cooperate and in doing so bring about the business of an enterprise. The nature of all the activities in an enterprise is either a coordination activity, by doing which subjects enter into coordination and comply with commitments, or a production activity, thus subjects contribute to bringing about the functions of an organization. Comparing to traditional process based on modeling methods DEMO is more suitable for describing complex social systems.

- DEMO has been theoretically proved to be able to support the design and simulation of Discrete Event System in several researches. DEMO is founded on δ-theory (DELTA theory, standing for discrete event in linear time automata) [16], which provides the basis for an appropriate understanding of what is commonly referred "how state of system is changed by event in a process". The other foundation theory of DEMO is the β-theory (β is pronounced as BETA, standing for binding (constructional) essence, technology, and architecture)[16] about the design of (discrete event) systems. It provides the base of an appropriate understanding of what is commonly referred to by engineering concepts that "separate design of the system development and implementation".

- DEMO has been practically proved to be able to support Discrete Event System in several researches. There are several researches[4][5][7][17]–[19] have argued DEMO as a conceptual model for simulation and contributed the ways of simulation with Petri net. In addition, a few studies [9] argued DEMO for discrete event simulation. These previous studies presented the importance and urgency of integrating enterprise modeling methods and executable simulation models in order to provide a better solution for design and analysis in the enterprise.

### 2.3 MDD4MS Framework

The MDD4MS framework defines the methodology for model driven development of simulation models through model transformations[20],following OMG's Model Driven Architecture (MDA)[21]. Table 1 demonstrates three levels of models: (1) Conceptual Model (CM); (2) Platform Independent Simulation Model (PISM); and (3) Platform Specific Simulation Model (PSSM). The meta-model of each layer should be an instance of higher level meta-model[11].

**Table 1.** Models and meta-models in MDD4MS framework[11]

| MDA Model | MDD4MS Model | MDD4MS Meta-model |
|---|---|---|
| Computation Independent Model (CIM) | Simulation Conceptual Model (CM) | Conceptual Modeling Meta-model |
| Platform Independent Model (PIM) | Platform Independent Simulation Model (PISM) | Model Specification Meta-model |
| Platform Specific Model (PSM) | Platform Specific Simulation Model (PSSM) | Model Implementation Meta-model |

## 3 DEMO Based DEVS Simulation Framework

Following three level MDD4MS, this research synthesizes a research framework in Fig 1. In this framework, we used DEMO as the conceptual model (CM layer); DEVS specification as the platform independent model (PISM layer); and DEVSDSOL, a Java based open source discrete event simulation platform, as the platform specified model (PSM layer). There are four meta-models defined in Generic Eclipse Modeling System (GEMS), including: ATD meta-model, PSD meta-model, AM meta-model and DEVS Step one (DEVS S1); Then four modeling tools are developed using Eclipse Modeling Framework (EMF), including: ATD modeling tool, PSD modeling Tool, AM modeling tool and DEVS S1 modeling tool; Models developed in the modeling tool could be translated into other model according to predefined transformation rules. The model transformation rules are defined using Eclipse ATL, including four model transformations: from ATD to PSD, from PSD to AM, from AM to DEVS S1, and from DEVS S1 to DEVS S2. The transformation from DEVS to DEVSDSOL is omitted in this paper since we use the existing research [11].

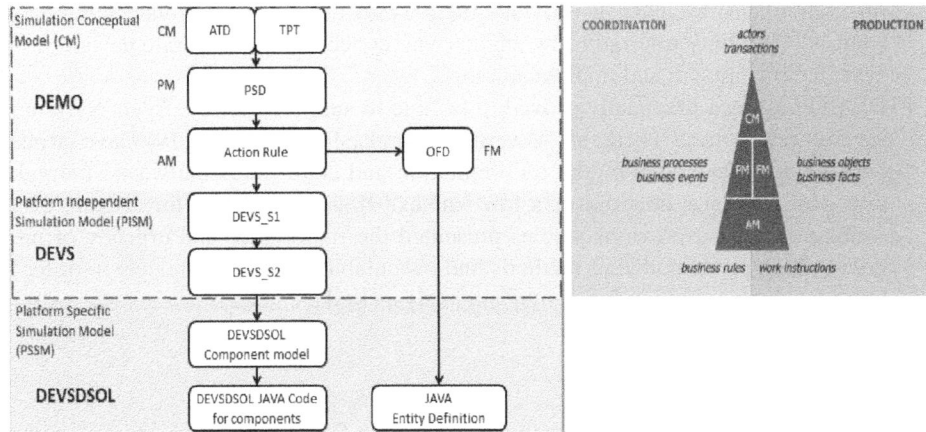

**Fig. 1.** Research Framework     **Fig. 2.** DEMO Aspect Models

### 3.1 DEMO Aspect Models

DEMO includes four aspect models that show different views of the completed ontological model.

As Fig 2 presented, construction model (CM) located on the top of the triangle, is the most concise model. In CM, transaction patens and corresponding actor roles are defined in ATD. Product of each transaction is given in transaction product table (TPT). The Action Model (AM) model, located on the bottom of the triangle, is the most comprehensive one. AM consists of action rules that specify every internal actor roles. AM also defines work instructions, including the execution of production act and judgment and decision of communication acts. Process model (PM) and Fact

Model (FM) are located in between CM and AM, for they are more detail than CM and less detail than AM. PM describes the coordination world from state view and process view, well FM describes the same thing for production world[22].

AM is used as conceptual model for simulation in this research. However, since it is not easy to get AM model directly, we generate AM step by step, from the most concise model CM to more detailed PM and finally get the most comprehensive AM. FM is used for defining entities used in AM.

### 3.2 Meta-models Definition

Following the definitions in DEMO, meta-models of CM, PM, AM and DEVS are defined in object entity diagram in GEMS platform.

In meta-model of CM (Fig 3): *AtdDiagram* represents the Actor Transaction diagram (ATD) in CM. The main graph element of ATD is defined as *AtdComponent*. There are two types of *AtdComponents*: *TransactionType* and *ActorRole*. *TransactionType* is the type of transaction pattern between actor roles. Each transaction has a related product defined in transaction product table (TPT). Here, *Product* is designed as one property of *TransactionType*. Since each transaction is related with one or more than one objects, *RelatedObj1* and *RelatedObj2* are also defined as properties. *ActorRole* is specified in two types: *CompositeActorRole* and *ElementaryActorRole*. The *CompoisiteActorRole* could be composed by *TransactionType* and *ElementaryActorRole*. *ActorRole* could be either initiator or executor of a *TransactionType*, thus there exist two types of link between *ActorRole* and *TransactionType*: *InitionLink* and Exe*cutionLink*.

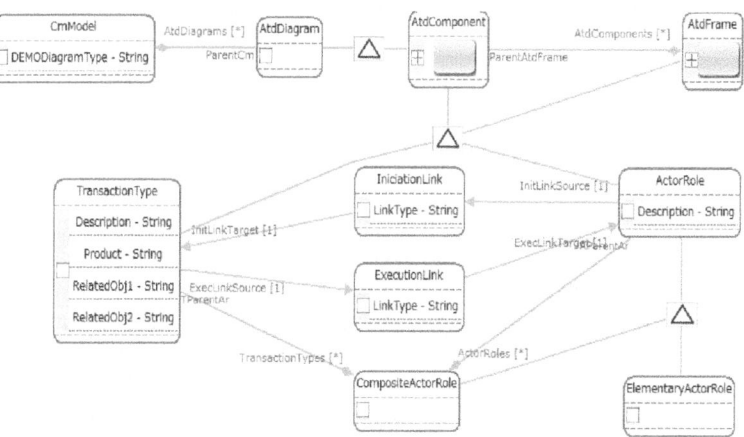

**Fig. 3.** ATD meta-model and XMI Expression

In meta-model of PM (Fig 4): *PsdDiagram* represents the Process Structure Diagram (PSD) in PM. The main element of PSD is defined as *PsdComponent*, which is specified as *ActorRole*. The element of *ActorRole* is defined as *ArComponent*, which is specified as *Act or INIT*. *Act* representing the responsibilities that an actor role took in communication and production process. There are two types of *Act*: *Cact*, the communication act, and *Pact*, the production act. For each act, the related object and product are derived from ATD. If there are any new objects or products generated by the act, it should be added to its property column. *INIT* represents starting point, where arrival rate and exit acts could be defined. There are two types of links between *ArComponents*: *CondLink*, representing conditional link, and *CausLink*, representing causal link.

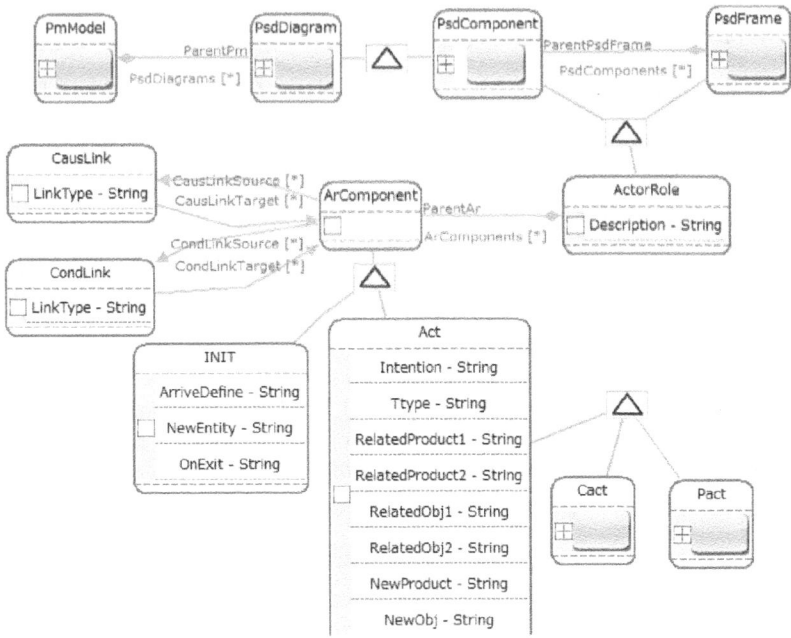

**Fig. 4.** PSD meta-model

In meta-model of AM (Fig 5): *AmDiagram* represents the Action Rule Diagram in AM. The main graph element of AM is *AmComponent*, specified as *ActorRole*. *ArComponents* is composed of *Act* and *INIT*, as defined in PM model. According to DEMO AM, each *ArComponents* is composed of: *WhenBlock* and *ThenBlock*. *WhenBlock* describes condition of an *ArComponent*, including properties related with current state of the world. *ThenBlock* is the judgment, the decision making and possible reaction of an *ArComponent*. It includes properties related with reaction and condition of the reaction. As Shown in Fig 5, *ThenBlock* of an *ArComponents* is linked to *WhenBlock* of another *ArComponents* by *CompLink* (component link), if the two *ArComponents* are linked in PSD model.

To generate complete simulation model, besides what has been defined in DEMO AM, additional information of resource is required. Resource could be seized or released by act. Whether the act uses a new resource, whether it seizes or releases the resources and whether there is a waiting queue required for the act defined by the properties of AM *Act*.

Automatic Model Transformation for Enterprise Simulation    143

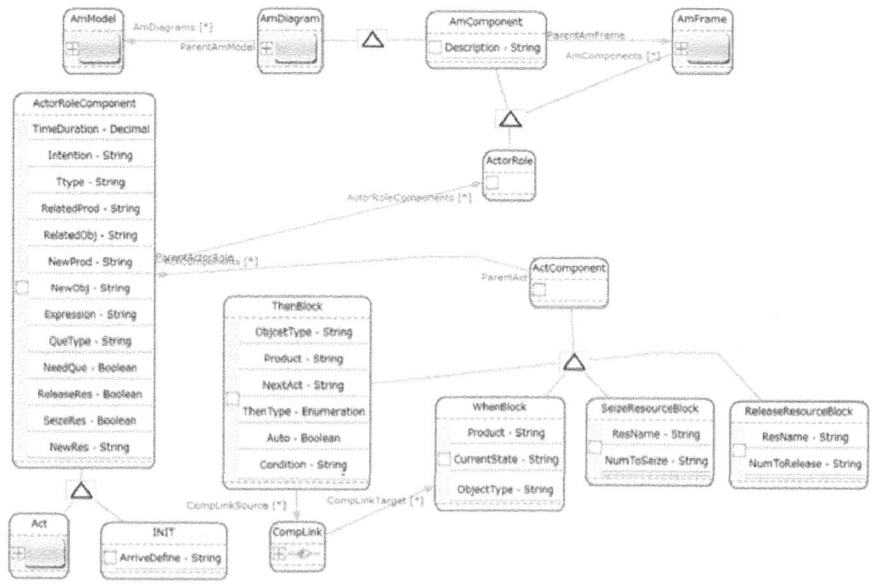

**Fig. 5.** AM/Resource meta-model

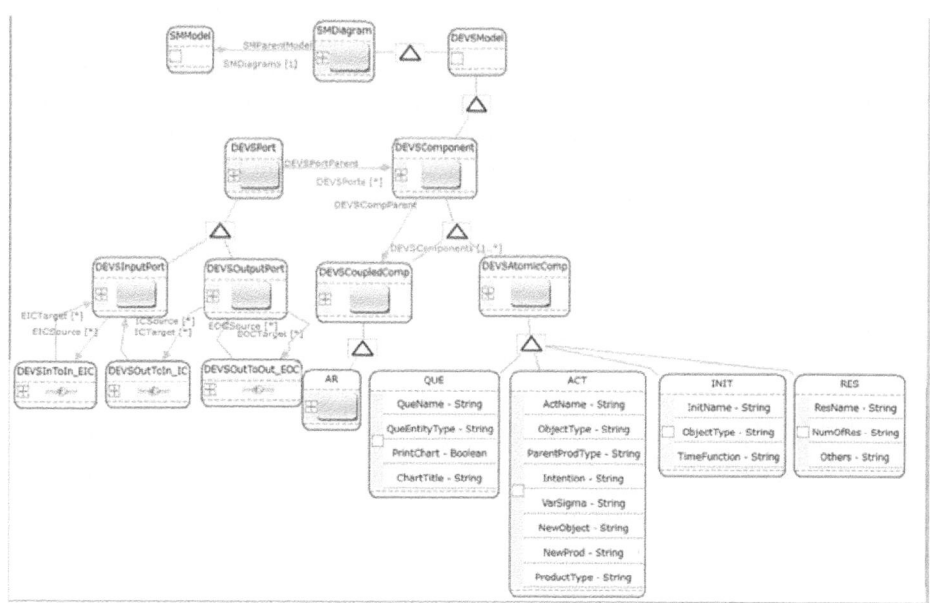

**Fig. 6.** DEVS S1 meta-model

In meta-model of DEVS S1(Fig 6), following the meta-model of DEVS [11], definition of InputPort, OutputPort, Links between inputPort and OutputPort, and components are given. There are five types of components defined in DEVS S1, including:

- AR (Actor Role), composite component, represents Actor Roles in DEMO;
- INIT (Initiation), atomic component, represents initiation points where new entity is created and the entity arrival rate is assigned;
- ACT (Act), atomic component, represents for Act, including both c-act and p-act in DEMO, where time duration of the act is defined
- Que (Queue), atomic component, represents for waiting queue of act. Queue could wait for resource, fact, or both
- Res (Resource), atomic component, represents resources. Resources are seized by waiting queue and released by an act.

Each type of components has its own DEVS specifications, which will be generated in transformation from DEVS S1 to DEVS S2. meta-model of DEVS S2 has been given [11] that we use the same definition in this research.

### 3.3 Model Transformation

After meta-model defined, four modeling platform is developed in GEMS framework, including: DEMO ATD modeling platform, DEMO PSD modeling platform, DEMO AM modeling platform and DEVS S1 modeling platform. Given transformation rules, the models in these modeling platforms, following the definition of its meta-model, could be semi-automatically generated through model transformation. Transformation rules define the mapping from source meta-model to target meta-model[10], [11] by using ATL (ATLAS transformation Language), the most popular model to model transformation language[23]. Four transformation rules are defined to finish the mapping from DEMO ATD to DEVS specification, including: T1: from DEMO ATD to DEMO PSD; T2: from DEMO PSD to DEMO AM; T3: from DEMO AM to DEVS S1 and T4: from DEVS S1 to DEVS S2 (left part of Fig 7). A sample code of ATL is presented in right part of Fig 7.

```
rule DEMOCActToDEVSAct (
    from
        s: AMMetamodel!Act
    to
        t: DEVSoneMetamodel!ACT  (
            DEVSCompParent <- s.ParentActorRole,
            Id <- s.Id + 1,
            Name <- s.Name,
            X <- 2*s.X,
            Y <- 2*s.Y,
            Width <- s.Width+50,
            Height <- s.Height+50,
            ExpandedWidth <- 2*s.Width,
            ExpandedHeight <- 2*s.Height+200,
            Expanded <- false,
            DEVSComponentType <- s.Name,
            Intention <- s.Intention,
            AtomicCompType <- 'ACT',
            NewObject <- s.NewObj,
            NewProd <- s.NewProd,
            ObjectType <- s.RelatedObj,
            ProductType <- s.RelatedProd,
            ParentProdType <- s.RelatedObj,
            VarSigma <- s.TimeDuration.toString()
        )
}
```

**Fig. 7.** ATL Sample Code

Transformation details are explained in Fig 8.

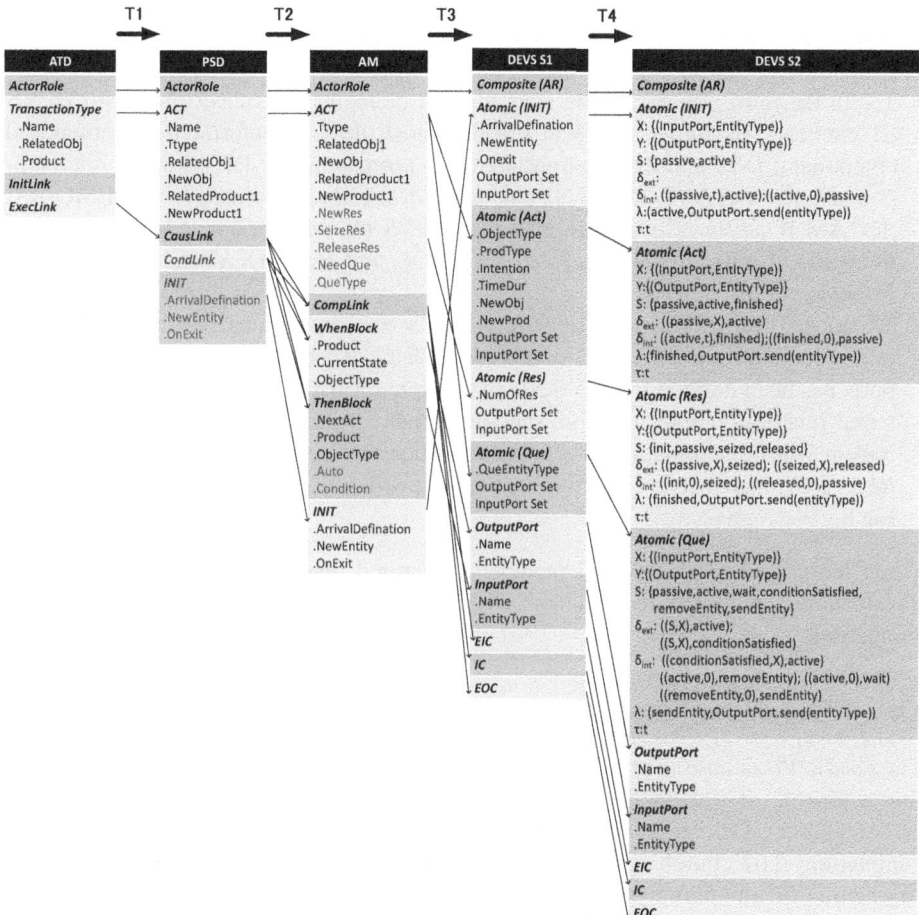

**Fig. 8.** Model Transformation

- T1: In this first transformation step, all ATD elements: *ActorRole, Transaction-Type* are transformed into PSD elements: *ActorRole, CAct, Pact*. The *ExecLink* are transformed into *CausLink*. Propertities of *TransactionType* are transformed into corresponding properties of act, as presented in T1 in Fig 8. After PSD model is automatically generated, we only need to add condition link, which describes the conditional constraint among acts, and then add *INIT* manually as the starting point to complete the PSD model (right color items in PSD in Fig 8).
- T2: In this second transformation step, as shown in T2 in Fig 8: All PSD elements are transformed into AM elements: *CondLink* and *CausLink* are transformed into AM *CompLink*, with a *WhenBlock* as the target of the link and a *ThenBlock* as the source of the link; Properties of *Act* in PSD are transformed into *Properties* of

146    Y. Liu and J. Iijima

*Act* in AM. After AM model is generated, condition specification in *WhenBlock*, and decision specifications in *ThenBlock* need to be manually given. Additionally, all the defination about resource needs to be manually added after the transformation to finish AM model (all the red color items in AM in Fig 8 need to be manually added).

- T3: In the third transformation step, all AM elements are transformed into DEVS S1 elements, as shown in T3 in Fig 8: ActorRole is transformed into composed component; *INIT* and *Act* are translated into corresponding DEVS S1 atomic components, *Act* and *INIT*. DEVS S1 *Que* and *Res* are also generated from DEMO AM *Act*. AM *WhenBlock* is transformed into DEVS *OutputPort* and AM *ThenBlock* is transformed into DEVS *InputPort*. *EIC*, *EOC* and *IC* in DEVS S1 are generated from *CompLink*.
- T4: In the fourth transformation step, DEVS S1 components will be transformed into corresponding DEVS S2 components. EIC, IC, EOC, input ports and output ports has been defined in DEVS S1. In the transformation T4, external function, internal function, output function, state and other formulations for each atomic component will be automatically generated. Notice that atomic component *INIT, ACT, REC, QUE* has different function and state specifications, as shown in DEVS S2 in Fig 8.

When the completed DEVS S2 model is generated, the work in [11] is utilized to transform the platform independent DEVS model into DEVSDSOL platform for simulation.

### 3.4    Case Study of Model Transformation and Simulation Result

The classic Pizza case is utilized for validation. In Pizza Case ATD model: There are four actor roles defined: CA01: Customer; A01: Order completer; A02; Order preparer; A03 Order Deliver. And four transactions between actor roles: T01: Purchase completion; T02: Purchase preparation; T03: Purchase deliver and T04: Purchase payment. Corresponding production for each transaction is defined in its property. ATD is draw in developed DEMO ATD modeling platform, and then semi-automatically transformed into PSD, AM, DEVS S1, DEVS S2 and finally executable DEVSDSOL Java code.

The model transformation process is expressed in Fig 9. ATD model is transformed into PSD model. Conditional link from T02ac to T03rq and conditional link from T03ac to T04pm and conditional link from T04ac to T01ex is manually added. The transformation from red rectangle part of PSD into AM is presented. Condition the reaction, if there is any, need to be manually added. About the resource, there is one oven and two stuffs in the store, one for accept order, the other one for prepare the pizza and deliver to customer. Customer coming rate is defined as a random *Exponential distribution* with mean value equal to 8. Normally it takes around 3 minutes to accept the order and 8 minutes for baking a pizza; deliver time is around 10 minutes in average. The information is assigned to corresponding properties of act in AM. The transformation from red rectangle part of AM into DEVS S1 is presented. The automatically generated DEVS specification of Act T01rq is shown in DEVS S2.

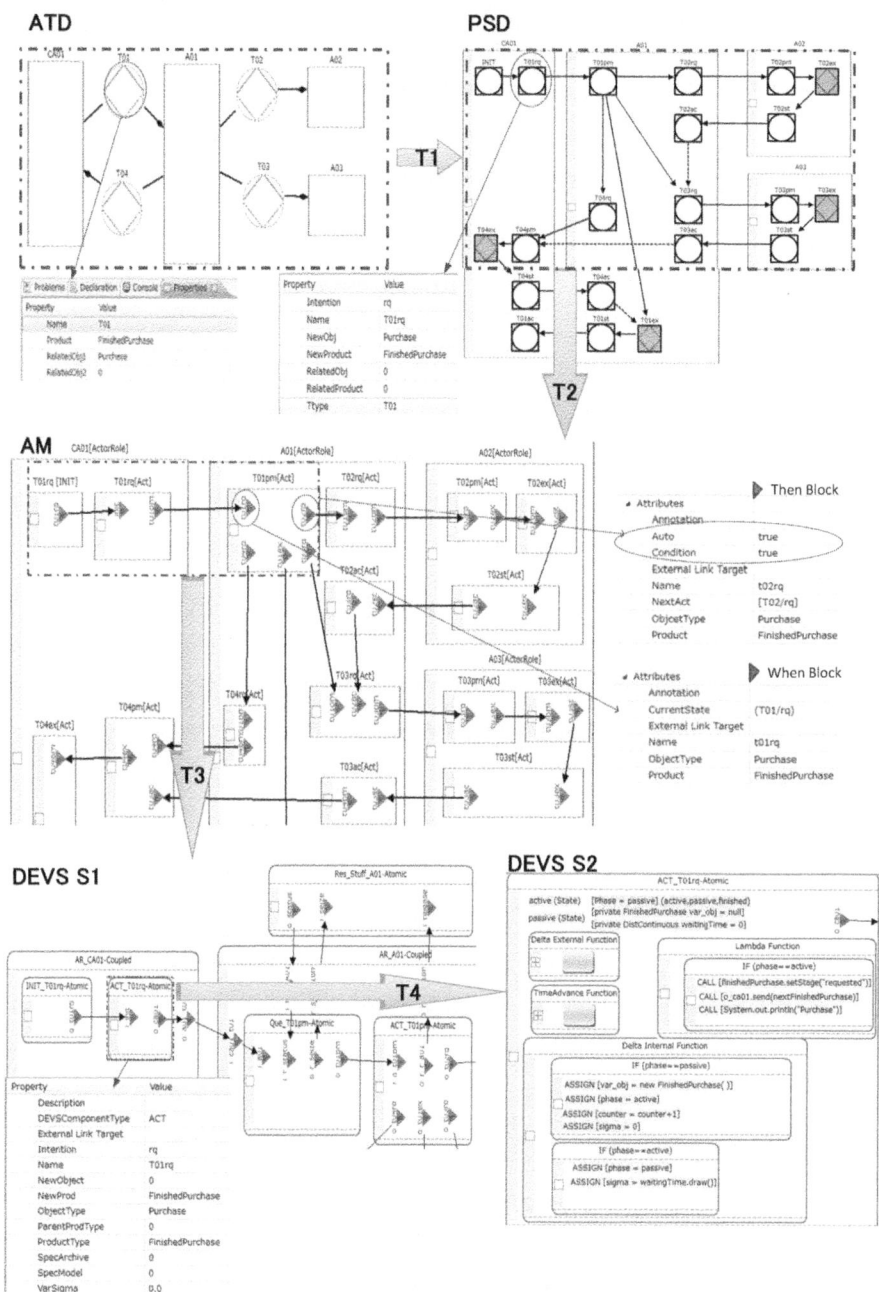

**Fig. 9.** Model transformation Sample

From DEVS S2, executable DEVSDSOL JAVA files is generated, as listed in Fig 10. Because of the complexity of simulation platform-DEVSDSOL and research limitation, the entities still need to be manually programed. Simulation environment and result are shown in figure 11. By using DEMO based simulation, the resource utilization, the bottleneck in the process could be analyzed.

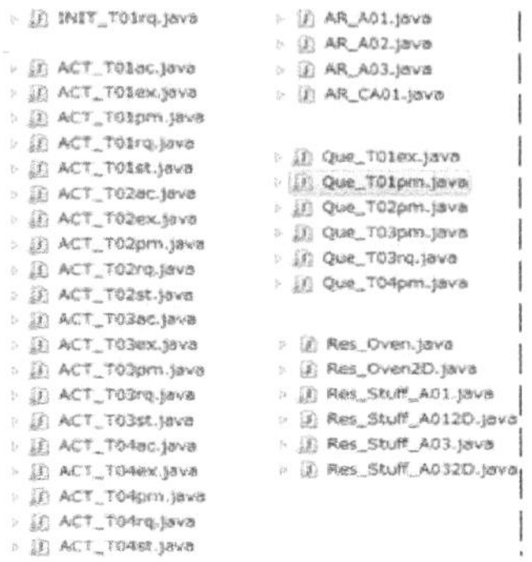

Fig. 10. Generated JAVA Files

```java
protected void deltaExternal(double e, Object input) {
    // TODO Auto-generated block
    if (activePort == i_init_t01rq) { //
        var_parentObj = (Obj_Purchase) input;
        phase = active;
        sigma = var_sigma;
        counter = var_parentObj.getID();
    }
}
@Override
protected void deltaInternal() {
    // TODO Auto-generated block
    if (phase == active) {
        try {
            time = this.simulator.getSimulatorTime();
            var_obj = new Prod_FinishedPurchase(this.simulator, counter,
                    objName, time, stageName, this, var_parentObj);
            var_obj.setParentEntity(var_parentObj);
        } catch (RemoteException e) {
            // TODO Auto-generated catch block
            e.printStackTrace();
        }
        phase = finished;
        sigma = 0;
    } else if (phase == finished) {
        phase = passive;
        sigma = Double.POSITIVE_INFINITY;
    }
```

Fig. 11. Generated JAVA Code

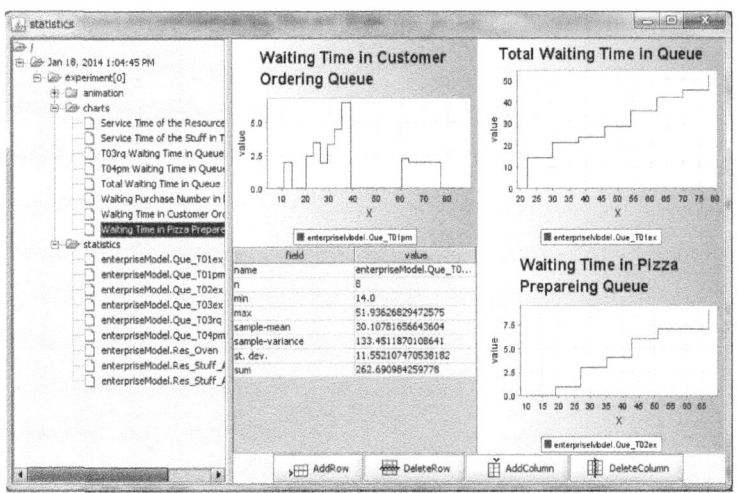

Fig. 12. Simulation Result

## 4 Discussion and Future Research

In this paper, we propose a methodology, which semi-automatically generate executable DEVS simulation from enterprise ontology. Based on this research, we summarize the following contributions for the community of enterprise researchers and practitioners:

First, it can help DEVS simulation developers who want to make simulation for an enterprise. The analysis and conceptual modeling begin by a concise construction level, and then go into detailed process, action rules, simulations related factors, resources, time duration, and finally get the executable DEVS simulation model. The whole process is semi-automatically completed that it prevent from unnecessary time consuming. Second, for enterprise engineering researchers, this study provides an approach for combining the modeling world with the simulation world in order to make a model executable. Third, in this study, we used JAVA to simulate the platform, and this research requires a lot of additional coding works manually after the code generated. However, since DEVS is the foundation of popular simulation tools such as Arena, or Anylogic. The generated DEVS S1 model could be easily transformed into other simulation platform. Thus this research could be expended to get make batter usage in practice.

The transformation from DEMO to DEVS in this research provides a possibility for semi-automatically generating executable models. However, this transformation still needs to be mathematically proved in our future research. Hence, we just use single server in this research, which will be improved by considering about the parallel situation. Also, DEVS entities defined according to DEMO FM are not automatically generated, which will be improved in the future research. In current stage we just consider about happy path in the communication loop. However, it is not always true in the real case. Consequently, the full path should also be focused on. In this research, we emphasize ontological constraint; however, in really simulation, we also need to consider the constraint in inforlogical level and information deliver method in the datalogical level. As we mentioned, the simulation platform DEVSDSOL is not easy for a normal user in their practice. Therefore, we will seek to transform generated DEVS S1 into other simulation platform for the better utilization in practice.

## References

1. Paul, R.J., Seranno, A.: Simulation for Business Processes and Information Systems Design. In: Proceedings of the 2003 Winter Simulation Conference (2003)
2. Gladwin, B., Tumay, K.: Modeling Business Processes with Simulation Tools. In: Proceedings of the 1994 Winter Simulation Conference, Lake Buena (1994)
3. Seila, A.F.: The case for a standard model description for process simulation. International Journal of Simulation and Process Modeling (2005)
4. Barjis, J.: Developing Executable Models of Business Systems. In: ICEIS 2007, pp. 5–13 (2007)
5. Barjis, J.: The importance of business process modeling in software systems design. Sci. Comput. Program. 71(1), 73–87 (2008)
6. Barjis, J., Dietz, J.L.G.: Business Process Modeling and Analysis Using Gert Networks. In: Enterprise Information Systems, pp. 71–80 (1998)
7. Barjis, J.: Automatic business process analysis and simulation based on DEMO. Enterp. Inf. Syst. 1(4), 365–381 (2007)
8. Wang, Y., Albani, A., Barjis, J.: Transformation of DEMO metamodel into XML schema. In: Albani, A., Dietz, J.L.G., Verelst, J. (eds.) EEWC 2011. LNBIP, vol. 79, pp. 46–60. Springer, Heidelberg (2011)

9. Liu, Y., Iijima, J.: A Research on Conceptual Model of Enterprise Simulation based on Enterprise Ontology (2013) (working paper)
10. Cetinkaya, D., Verbraeck, A., Seck, M.D.: A Metamodel and a DEVS Implementation for Component Based Hierarchical Simulation Modeling. In: Proceedings of the 2010 Spring Simulation Multiconference, SpringSim 2010, pp. 130–137 (2009)
11. Cetinkaya, D., Verbraeck, A., Seck, M.D.: Model Transformation from BPMN to DEVS in the MDD4MS Framework. In: Proceedings of the 2012 Symposium on Theory of Modeling and Simulation - DEVS Integrative M&S Symposium, TMS/DEVS 2012 (2012)
12. Zeigler, B.P., Kim, T.G., Praehofer, H.: Theory of Modeling and Simulation, 2nd edn. Academic Press (2000)
13. Dietz, J.L.G.: Enterprise Ontology. Springer (2006)
14. Lehtinen, K., Lyytinen, E.: Action Based Model of Information System. Inf. Syst. 11(4) (1986)
15. Goldkuhl, G.: Business Frameworks and Action Modelling. In: LAP (1996)
16. Dietz, J.L.G., Hoogervorst, J.A.P., Albani, A., Aveiro, D., Babkin, E., Barjis, J., Caetano, A., Huysmans, P., Iijima, J., Van Kervel, S.J.H., Mulder, H., Op't Land, M., Land, T., Proper, H.A., Sanz, J., Terlouw, L., Tribolet, J., Verelst, J., Winter, R.: The discipline of enterprise engineering. Int. J. Organ. Des. Eng. 3(1), 86 (2013)
17. Barjis, J., Dietz, J.L.G.: Language Based Requirements Engineering Combined with Petri Nets. IFIP AICT, pp. 1–12 (2000)
18. Barjis, J., Dietz, J.L.G., Liu, K.: Combining the Demo Methodology with Semiotic Methods in Business Process Modeling. In: Information, Organisation and Technology; Studies in Organizational Semantics, pp. 213–246 (2001)
19. Barjis, J., Kolfschoten, G.L., Verbraeck, A.: Collaborative Enterprise Modeling. In: Proper, E., Harmsen, F., Dietz, J.L.G. (eds.) PRET 2009. LNBIP, vol. 28, pp. 50–62. Springer, Heidelberg (2009)
20. Cetinkaya, D., Verbraeck, A., Seck, M.D.: MDD4MS: A Model Driven Development Framework for Modeling and Simulation. In: Proceedings of the 2011 Summer Computer Simulation Conference (2011)
21. Object Management Group, "Model Driven Architecture MDA" (2000)
22. Perinforma, A.P.C.: The essence of organisation (2012) ISBN: 978-90-815449-4-8. Sapio Enterprise Engineering
23. ATL, Specification of the ATL (Atlas Transformation Language) Virtual Machine Version 0.1 (2005)

# Introducing a Framework for Scalable Dynamic Process Discovery

David Redlich[1,2], Wasif Gilani[2], Thomas Molka[2,3], Marc Drobek[2,4], Awais Rashid[1], and Gordon Blair[1]

[1] Lancaster University, United Kingdom
mr.redlich@gmail.com, {gordon,marash}@comp.lancs.ac.uk
[2] SAP Research Center Belfast, United Kingdom
wasif.gilani@sap.com
[3] University of Manchester, United Kingdom
thomasmolka@gmail.com
[4] Queen's University Belfast, United Kingdom
MarcDrobek@gmx.de

**Abstract.** Businesses are becoming increasingly globally interconnected and need to continuously adapt to global market changes and trends in order to stay competitive. Business processes are fundamental parts and drivers of these globally connected organizations which is why their management, analysis, and optimization are of utmost importance. Discovering and understanding the actual execution flow of processes deployed in your organization is an important enabler for these tasks. However, this has become increasingly difficult since business processes are now mostly distributed over different systems, highly dynamic, and may produce thousands of events per second which may conform to a number of different formats. These particular challenges are currently not specifically accounted for in the research field of Process Discovery. In order to address these challenges, this paper presents a concept for scalable dynamic process discovery, which is a scalable solution for identifying and keeping up with the evolution of dynamic, collaborative business processes. Furthermore, a framework for this concept is proposed along with the requirements and implementation details for the involved components and models.

**Keywords:** Business Process Management, Process Discovery, Enterprise Architecture, Complex Event Processing.

## 1 Introduction

Due to globalization big organizations are facing a rising competition and have to become increasingly adaptive to market changes and are thus constantly in a process of optimization. At the core of these organizations are business processes which define the flow of work for high-level business functions that help to achieve important goals and are considered to be "...*the most valuable corporate asset*" [1]. In order to continuously optimize your organization and its deployed

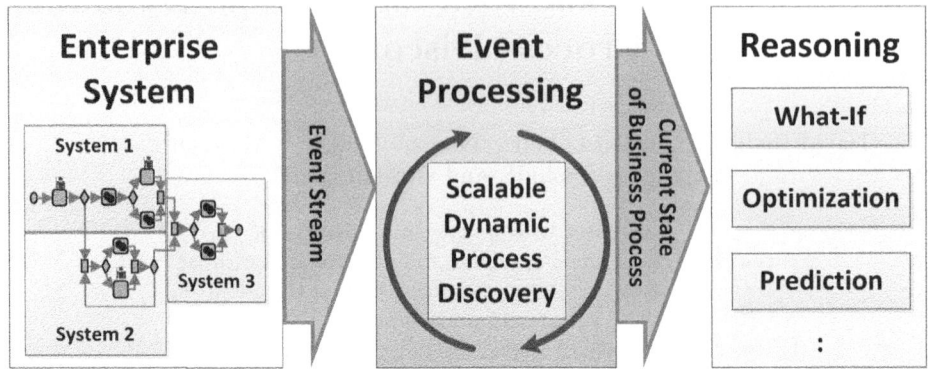

**Fig. 1.** Scalable Dynamic Process Discovery - Conceptual Overview

business processes you first need to have information about what is actually going on in your company right now, i.e. what is the current condition of your business processes. This information helps you make insightful decisions and/or answer questions like: How are my business functions executed in reality? How do my business processes and employees perform? What are bottlenecks of the current business processes? However, discovering and understanding the actual state of your deployed processes has become increasingly difficult. There are many different reasons for this of which the most influential are:

- business processes often span multiple different systems and thus produce different types of events during execution respectively,
- business processes are more frequently changing over time, i.e. became more dynamic, and
- the execution of core business processes of large organizations are likely to produce hundreds of events per second.

Up until now these particular challenges are usually not considered in the research field of traditional Process Mining. In this paper we propose the concept of Scalable Dynamic Process Discovery (SDPD) that addresses these challenges. SDPD is a scalable solution for identifying and tracking the evolution of dynamically changing collaborative business processes. Employing this concept allows for processing events from different sources at run-time to gain information reflecting the current state of the monitored business process. This allows for high-level analyses based on the discovered state, such as what-if analysis, prediction, and optimization. Figure 1 shows the general concept of Scalable Dynamic Process Discovery. One important feature of this concept is that it is independent from any additional model input, i.e. the important aspects of a process like control-flow, performance, etc. are mined during run-time and are, as defined in [14], part of the "state" of the process. This poses a particular challenge for discovering the control-flow of a process at run-time: To the best of the authors' knowledge all known control-flow discovery algorithms are not intended to be used in run-time event processing. Along with the definition of Scalable Dynamic Process Discovery a framework is described which has been developed as an implementation of the concept. Said framework is a result of and driven by the requirements of real life industrial use cases[1] provided by business partners within the EU funded project TIMBUS[2].

The content of this paper is structured as follows: First, background information about research topics related to the SDPD concept is provided. Then in Section 3, Scalable Dynamic Process Discovery is introduced via definition plus characteristics and additional requirements are specified. In Section 4, a framework that addresses the identified challenges of SDPD is presented along with the description of all involved agents and models. In the last Section 5 the paper is summarized and future extensions and improvements are discussed.

## 2 Background

In an organization elementary tasks have to be carried out in a certain way in order to achieve business goals and meet predefined objectives. These tasks, their order of execution, and the resources to perform them are usually modelled in one or more business processes. Ko et al. define business processes in [8] as *"...a series or network of value-added activities, performed by their relevant roles or collaborators, to purposefully achieve the common business goal."* Well known examples of such processes are Order-to-Cash, Accounts-Receivable, or Procure-to-Pay. The listed examples are technically not business processes but *process types*, i.e. they are of a specified type of process with a defined business goal. A business process is usually represented by a *process model* specifying the aspects of the business process, e.g. involved activities and resources, plus the execution order. Popular example standards for process models are Business Process Model and Notation (BPMN) [12], Event-driven Process Chains (EPC) [15], and Yet Another Workflow Language (YAWL) [19]. A process type may be represented by different process models, each expressing another version or evolution step of this process type. A *process instance*, on the other hand, is defined as a single execution of the business process; it is also referred to as *trace* or *case* in some literature. Furthermore, it is common in business process terminology to distinguish between different perspectives of a business process. In the context of this paper we take the following perspectives into account:

- Control-flow: This perspective describes the execution order of the single activities with the help of control elements, e.g. parallel splits and decisions. This perspective is mostly referred to when using the term business process.
- Resources: This perspective describes the resources that are in charge of executing each activity in the control-flow.

---

[1] In the domain of eHealth and dam safety.
[2] TIMBUS is about preserving business processes and in this context it is desirable to know at which point in time a process has changed.

– Performance: This perspective describes the performance of a business process, which includes for instance information about the execution time of an activity, or how often the process has been initiated, or with which probability certain paths are chosen.

Not listed above is the data perspective which is concerned with specific information that is associated with a process instance, e.g. items to be shipped, money to be transferred. Unfortunately, the data perspective is highly dependent on the process specifics and implementation of the process and is therefore not easily generalized. Since this paper focuses on a general solution that is independent from external input and as such not use-case specific, we focus on the perspectives that can be generalized: Control-flow, Resource, and Performance.

The increasing complexity and importance of business processes initiated the development of Business Process Management (BPM) as an IT-related research area during the last decade [8]. In fact, BPM itself is a cross-discipline subject of "theory in practice" adopting a variety of concepts and methodologies, e.g. computer science, management theory, philosophy, and mathematics [8]. According to van der Aalst BPM is defined as follows: *"Supporting business processes using methods, techniques, and software to design, enact, control, and analyze operational processes involving humans, organizations, applications, documents, and other sources of information."* [17]. Software systems that support the execution and general management of operational business processes are referred to as Business Process Management Systems or Business Process Management Suites (BPMS's) [9]. Popular examples of BPMS are SAP Netweaver BPM [22] or Intalio BPMS Designer [6].

When business processes are executed in BPMSs they produce a *Log* which is a record of occurring events. These logs can look very different since every BPMS may have its own format. Apart from the format the produced log can also differ in other aspects, e.g. the event granularity or contained information. Examples of event formats are XES [5] and BPAF [23]. In contrast to the approach of storing logs and afterwards analysing them stands the method of immediately processing these events when they occur in order to enable real-time analysis. This is achieved with the help of Complex Event Processing (CEP), which is a method that essentially deals with the event-driven behaviour of large, distributed enterprise systems [10]. This means in particular that events produced by the systems are captured, filtered, aggregated, and eventually abstracted to generate complex events representing high-level information about the situational status of the system. The need for continuously analysing a business process by applying CEP methodologies has been identified by Ammon et al. who coined the term Event-Driven Business Process Management (EDBPM) [1]. The term emerged from the combination of the two disciplines Business Process Management and Complex Event Processing [1]. This is practically realised by two individual platforms interacting with each other through interfaces or events: One is a BPM system, which is used to model, manage, and optimise a business; the other one is a CEP engine [2]. If a CEP engine is configured to compute real-time information about the performance of a business process (see performance perspective) it is called Business Activity Monitoring (BAM). Single live-events are not of interest in the context of BAM, instead the aggregation of these into performance related parameters is carried out [3]. BAM solutions, e.g. [7,13,4], are per definition applications of EDBPM.

Another very important part this paper is concerned with is the topic of process mining, a research discipline that is located at the intersections between machine learning, data mining, process modelling, and process analysis [20]. Process Mining describes the method of discovering, monitoring, and improving real processes using knowledge extracted from an event log produced by actually executed processes. Three main disciplines of process mining exist: (1) *conformance* - comparing an existing process model with an event log of this process, (2) *enhancement* - extending an existing model with additional information obtained from the event log of this process, and (3) *discovery*. Process discovery is concerned with extracting a business process model from an event log without using any a-priori information [20]. BAM solutions are in the context of process mining usually classified as enhancement, since basic information about the process is already provided, e.g. [13].

The challenges of process discovery are generally motivated by the accuracy and quality of the result, e.g. precision, simplicity, fitness - over-fitting vs. underfitting (see [20]). Of less or little importance on the other hand is the practical execution of these process discovery solutions during run-time: As stated in the previous paragraph it is a static method that analyses a complete event log in order to find the most accurate business process model conforming to the input event log. This fact is reflected in the following definition:

**Definition 1.** *Let the log $L_n = [e_0, e_1, ... e_n]$ be a sequence of $n+1$ events ordered by time of occurrence ( $\forall i < j \land e_i, e_j \in L_n : time(e_i) < time(e_j)$) and $BP_n$ be the business process model representing this sequence of $n+1$ events then process discovery is defined as a function that projects log $L_n$ to $BP_n$, i.e.*

$$ProcessDiscovery : (e_0, e_1, ..., e_n) \rightarrow BP_n$$

## 3 Scalable Dynamic Process Discovery

As established in the previous section extensive research is being carried out in the areas of Business Activity Monitoring, Event-driven Business Process Management, and Process Discovery - just to name the few most relevant areas for this topic. However, due to increasingly volatile and collaborative processes as well as the need for getting immediate insight into your business, current research is driven by a new set of challenges. To address these problems we introduce the concept of Scalable Dynamic Process Discovery (SDPD), an interdisciplinary concept employing principles of CEP, BAM, Process Discovery, and EDBPM.

SDPD describes the method of monitoring one or more BPMSs in order to provide at any point in time a reasonably accurate representation of the current state of the processes deployed in the systems with regards to their controlflow, resource, and performance perspectives as well as the state of still open traces. That means, any change in the mentioned aspects of processes in the system during run-time has to be reflected in the monitored representation of the current state. This definition results in a set of special characteristics and additional requirements that SDPD needs to comply to:

- Extensibility: Since SDPD allows for monitoring processes spanning multiple different BPMSs it is necessary that it can deal with the different formats of the events produced by the BPMSs. That is why SDPD should enable the introduction of additional adapters which allow for processing new event formats. Furthermore the individual monitoring and reasoning components should be interchangeable.
- Detection of Change: SDPD should detect change in the two different levels defined in [14]: (1) Reflectivity: A change in a process instance (trace), i.e. every single event leads to a change in the state of a trace. (2) Dynamism is a change on the business process level, i.e. if the recent events indicate a change of one of the perspectives of a business process, e.g. because a trace appeared that contradicts with the assumed control-flow.
- Scalability/Algorithmic Run-time: With regards to the control-flow discovery of processes this was previously of almost no importance but since SDPD is applied as CEP concept and has to deal with potentially massive business processes consisting of hundreds of activities, the actual run-time of the deployed algorithms becomes very important. The SDPD concept requires scalability in order to cope with increasing workload at as little as possible additional computational cost.
- Generalization/Standardization: As discussed in the background section, many business process representations exist. However this concept is required to work for a standardized model which supports the most common elements of the existing standards of the business process domain, but not special elements that are only supported by a minority of the standards.
- Accuracy: Accuracy is always compromised when achieving the goal of building up a general purpose solution and on top of that meet additional algorithmic run-time constraints. The accuracy of the SDPD could therefore be lower compared to a specialized solution, but still can be increased by applying certain customizations such as developing target specific adapters, etc.

Driven by these challenges the initial concept of process discovery has to be altered in order to allow for dynamic process discovery. Instead of the traditional static method (see Definition 1) dynamic process discovery is an iterative approach as defined in the following:

**Definition 2.** *Let the log $L_n = [e_0, e_1, ...e_n]$ be a sequence of $n+1$ events ordered by time of occurrence ( $\forall i < j \land e_i, e_j \in L_n : time(e_i) < time(e_j)$) and $BP_n$ be the business process model representing this sequence of $n+1$ events then dynamic process discovery is defined as a function that projects the tuple $(e_n, BP_{n-1})$ to $BP_n$, i.e.*

$$DynamicProcessDiscovery : (e_n, BP_{n-1}) \to BP_n$$

## 4 A Scalable Dynamic Process Discovery Framework

According to the definition of SDPD we have created a framework that complies to the characteristics and requirements listed in the previous section: Different BPMSs provide the input in the form of events and the result provided by the framework is the complete current state of these monitored systems, including the control-flow, resource, and performance perspective as well as the current state of the active traces. However, with regards to the overall concept there were two general challenges that had to be addressed:

1. The first one was to create or extend existing approaches in the area of control-flow discovery to make them work in a scalable manner, i.e. the runtime of the processing of a single event is required to grow at most linearly with regards to the number of activities involved in the process and has to be independent from the total number of events received.
2. Another challenge was to make the monitoring of the state of traces as well as the performance perspective independent from the structural information given in the control-flow and resource perspective. The control-flow information was needed, for instance, if path probabilities for decisions of a given business process were to be monitored.

In order to address these challenges the concept was divided into two parts: (1) The *Event Processing* operating at run-time, complying to the requirements of scalability, and producing a so called *dynamic footprint* of the event input; and (2) the *Footprint Interpretation* which can extract the actual state of the business process based on the current dynamic footprint. The footprint interpretation has less restrictions with regards to the scalability requirement as it does not have to be executed with every occurring event but rather more autonomously, i.e. either on demand, or repeatedly after a certain time has passed or after a fixed number of events or traces occurred.

The resulting conceptual framework is presented in an information flow diagram in Figure 2. It shows agents in a rectangular shape and models with round edges. Note that to improve the understanding for the reader the concept depicted and explained focuses on the monitoring of one end-to-end process only.

The general concept works as follows: Events from different sources of the monitored *Enterprise System*, in which the end-to-end process is deployed, are processed to a standardized format and put into a global context by the *Event Hub*. The standardized events are then further processed to update the current dynamic footprint, which acts as the current state of the process. The footprint information can then at any point in time be compiled to the actual state information of the business process, i.e. the abstract footprint representation is interpreted into knowledge conforming to a generalized business process standard (control-flow, performance, and resource perspective, and state of the active traces). This information can be processed by different reasoning algorithms to further analyse the process, e.g. performance prediction via simulation [13].

In the following the models and agents involved in the SDPD framework are described in more detail. This includes additional requirements, further

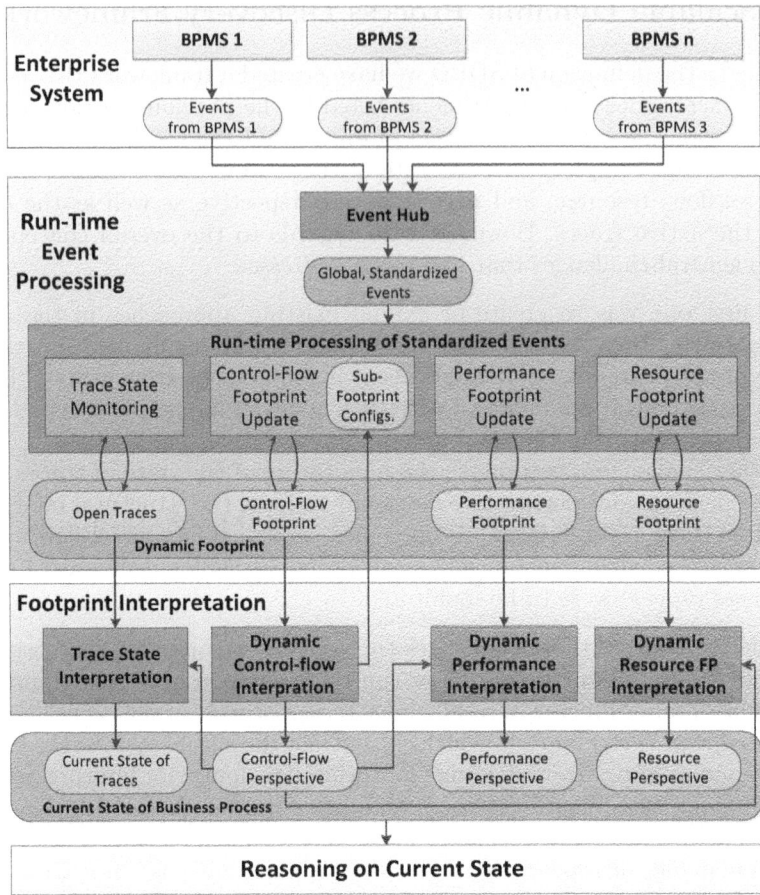

**Fig. 2.** Information Flow: Agents and Models involved in the SDPD Framework

specification, a few important implementation details and findings, and, if existent, references to similar and existing work associated with occurring challenges.

### 4.1 Event Hub and Global, Standardized Events

The *Event Hub* and the *Global, Standardized Events* are part of the framework mainly with the purpose of complying to the extensibility requirements of SDPD. They are responsible for the pre-processing step in which every event is translated into a standardized version that can be further processed.

*Global, Standardized Events.* The *Global, Standardized Events* are pre-processed events conforming to a general format. In the case of this framework the format used is inspired by and conforms mostly to the XES format [5]. However, as this is a run-time solution the notion of *Log* is not supported and the notion of a *Trace* is included in the attributes of an event. Generally, the following information needs to be present if a complete state of the business process as defined by SDPD is to be monitored:

- *Process ID* is necessary to map an occurring event to its business process. It is only required if multiple processes are monitored.
- *Trace ID* is the association of an event to a certain trace/process instance.
- *Process Element* is the reference to the process element associated with this event. It usually refers to process activities or process events (e.g. *Start, End* event).
- *Timestamp* is the time when the event occurred in the system (i.e. it is not the time the event hub received the event). It is required to monitor time-related performance information as well as order events according to their actual occurrence, if needed.
- *Lifecycle Transition* describes the lifecycle transition of an activity which caused this event. The XES standard proposes an extensive set of possible transitions that represent the lifecycle of an activity execution, however most BPMS only support a subset of these. In the proposed standardized format of this framework the lifecycle transitions *scheduled, assigned, completed* are considered.
- *Resource* is the reference to the entity that was responsible for performing/carrying out the activity.

Additionally to the events produced by executing an activity a special *End* event is needed to indicate that the life time of this trace is over. If an End event appears, that means the state of the trace turns from *open* into *closed*. Also, the conformance to the XES format was chosen to enable an extension of the format at a later point in order to allow for additional information to be mined, further supporting the extensibility requirement of SDPD.

*Event Hub.* The main task of the *Event Hub* is the translation of BPMS-specific events to events which conform to the standardized format. This is achieved via adapters which implement this translation. With regards to the translation the following problems had to be addressed: (1) Two BPMS might work in different timezones, which is why the timestamp value has to be translated to a unified timezone (e.g. UTC) in order to avoid analysis errors; (2) activity names might have different formats or might collide even though they originate from two different activities. In these cases a mapping to a unified naming system has to be in place for the various adapters. The same applies for other event information, e.g. the *Trace ID*: If two different BPMS use two unequal trace ids for the same trace, the event hub has to map them while processing, e.g. by using an alternative event feature to create the mapping.

### 4.2 Dynamic Footprint

The *Dynamic Footprint* (see Figure 2) is the model that contains the state of the business process in an abstract form and acts both as an input and output of the run-time event processing. With each event the footprint is potentially updated. It has to be noted at this point again that the dynamic footprint model is only abstract information and still has to be interpreted into a proper business process model. The term footprint has been borrowed from the process discovery terminology: In many process discovery algorithms it is common to first build a footprint which is then analysed and transformed into a business process model, e.g. alpha algorithm [16] and heuristics miner algorithm [21].

One of the main challenges for the framework was the design of the footprint: On one hand it has to be expressive enough to enable the translation into the different aspects of a business process state; on the other hand size constraints were to be met in order to ensure a quick update of the footprint during runtime. Furthermore, to support the scalability the size of the footprint has to be independent from the total number of occurred events and from the total number of occurred traces. This is necessary to keep the run-time at a constant value. Only the number of activities and resources is influencing the size of some parts of the dynamic footprint which is explained later. Another finding during the development of the framework was that, apart from the *Open Traces*, all parts of the footprint should avoid absolute statements, i.e. true/1 and false/0, but instead use weights, e.g. statement A is true with a probability of 0.92 on a scale from 0 to 1. These statement weights can than be updated incrementally with each event, either supporting or opposing the statement.

*Control-Flow Footprint* consists of three matrices: (1) *eventually follows* which is to capture the global relation between two activities, (2) *before first appearance* which is a relation matrix that helps identifying splits, and (3) *direct neighbours* that contains probability information about which activities usually directly follow a specified activity - its concept is very close to the footprint proposed for the heuristics miner-algorithm [21]. Each of the matrices have the size of $n * n$ with $n$ being the number of involved activities. This is necessary because each activity could be connected with all other activities, i.e. star-network [20]. Two more vectors exists storing the probability and the average count of an activity occurring in a trace. This information is for instance needed to identify loops.

*Performance Footprint* consists of generic performance parameters like "process instance occurrence" or "activity networking time" stored as a normal distribution function, i.e. mean and deviation values. The size of this footprint increases linearly with the amount of activities in the process. An exception is the path probabilities for decisions: For them the footprint is the *direct neighbours* matrix as introduced in the previous paragraph. If they are to be mined the size of the footprint increases quadratically in relation to the amount of involved activities.

*Resource Footprint* consists of a matrix that associates each activity to a resource with a certain weight.

*Open Traces* consists of the last lifecycle transition of each activity that has appeared in each open trace. This is the only part of the footprint that consists of absolute statements.

## 4.3 Current State of Business Process

The *Current State of Business Process* (see Figure 2) is the interpreted dynamic footprint into a business process notation conform model. It consists of the three perspectives of control-flow, resources, and performance plus the information about the current state of the traces. The current state of the business process acts as an input for known BPM reasoning techniques like simulation in order to perform for instance a prediction or what-if analysis. The main challenge for this model was to find a generalized representation that can be analysed with existing reasoning methods and at the same time supports the common element types that can be found in popular business process standards.

*Control-Flow Perspective* : While in industry Business Process Model and Notation (BPMN) [12], Business Process Execution Language (BPEL) [11], and Event-driven Process Chain (EPC) [15] are the most prominent examples, in research Yet Another Workflow Language (YAWL) [19] is considered to be the most established standard. This diversity of standards makes it difficult to determine one general standard. However, for this framework we focused on a general set of control-flow constructs that can be expressed by most standards:

- *Start and End Event* are basic constructs in a process model indicating the entry point of an instantiation (start event) or the exit point, i.e. a termination (end event) of a process instance.
- *Activity* is the actual work that has to be executed. It can either be atomic or has its own lifecycle. It is distinguished between human and automated activity, both of which have a different lifecycles, i.e. human activities posses the notion of a queue.
- *AND-Split/Join* are used to direct the work flow of the process. The XOR-Split represents a forking of the current instance into two or more parallel work flow paths. Its counterpart, the AND-Join, represents a synchronization point for the instance - it enables when all of the incoming paths are completed.
- *XOR-Split/Join* are also used to direct the work flow of the process. The XOR-Split is semantically equal to an exclusive decision for exactly one of the target paths. Its counterpart, the XOR-Join, is the unsynchronized merge element which is enabled once one of the incoming paths is completed.

Many BP standards may also support further high-level constructs but these can usually be reconstructed by a set of the mentioned low-level constructs. In Figure 3 an example process involving all the introduced elements is displayed.

*Resource Perspective* : This perspective contains information about which *resource* is associated to which *role(s)* and which activity is performed by a resource of a certain *role*. E.g. Activity "Pay Compensation" can only be performed by a resource of the role "Accountant" and "Tim" is a "Manager" and "Accountant", which means he is able to carry out the activity "Pay Compensation". Similarly to activities, two types of resources exist: human actors associated to human activities and machine actors associated to automated activities.

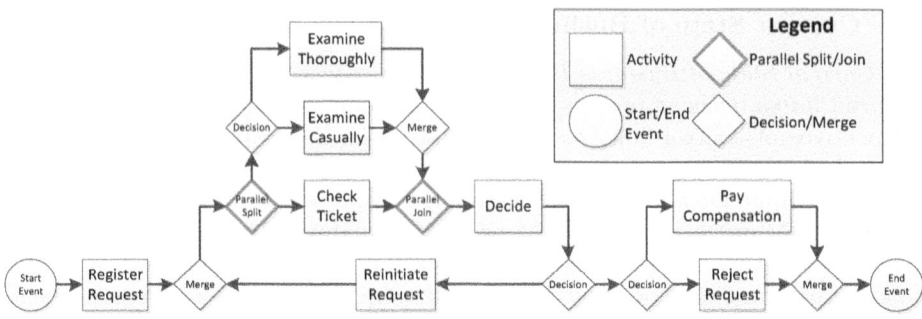

**Fig. 3.** Example business process with all element types included [18]

*Performance Perspective* : This perspective provides information about certain performance aspects of the business process very much like BAM solutions offer, e.g. [7,13]. For the framework the following general performance parameters are part of the state: *Process Instance Occurrence, Activity Net Working Time, End-To-End Processing Time, Queue Length,* and *Decision Path Probabilities.*

*Current State of Traces* : Since the data perspective is too system/domain specific it cannot be monitored by the general purpose solution offered by this framework. Instead the state of the traces is captured. This information can be used for reasoning, e.g. as an initial state for a simulation.

### 4.4 Run-Time Processing of Standardized Events

As shown in Figure 2 the *Run-time Processing of Standardized Events* agent processes the global standardized events to updates in the dynamic footprint and thus updating the abstract state of the business process. Each of the agents included in this component has to be an independent deterministic method without any input apart from the events. The *Control-flow Footprint Update* agent is to some extent an exception to this constraint. Furthermore, each processing agent should only take a constant amount of time, independent from the total number of previously occurred events or traces. The total number of involved activities can have, however, a linear increase of the run-time due to recalculating the relation of the occurred activity to, in the worst case, all other activities. Overall the agent has to scale linearly to the number of events occurring. Since the event stream is to be processed sequentially due to the incremental methods applied the framework can be scaled up easily by adding computational power.

Essentially, the main challenge was to split up the process discovery algorithms into a footprint extraction part and an footprint interpretation part. Additionally, the footprint extraction part had be dynamic, i.e. work incrementally and always update the current dynamic footprint in a way that older events have less influence than the newer events. The approach utilized in the SDPD framework is to build for each trace its own trace footprint[3] ($TFP$) and at trace

---

[3] With absolute statements true=1, false=0.

end add this multiplied by a factor $p$ to the dynamic footprint ($DFP$) multiplied by $1 - p$, e.g. for $p = 0.01$: $DFP = 0.01 * TFP + 0.99 * DFP$. That means, a trace footprint $TFP_i$ has when added the influence of 0.01, after another $TFP_{i+1}$ has been added the influence of $TFP_i$ would be $0.01 * 0.99$, and after another $0.01 * 0.99^2$ and so on. Through applying this method older $TFP$ are losing influence in the overall dynamic footprint. Note that the Trace State Monitoring is not updated that way but instead only the lifecycle transition for the activity and trace associated with the occurred event is updated (see 4.2: Open Traces).

One fact not to be ignored is that if any information is not available from the events, some parts of the footprint cannot be discovered, e.g. if the resource information is not provided the resource footprint cannot be build, or if the lifecycle transitions were not to be provided, the performance footprint in general and the activity net working time in particular could not be properly discovered without further information about the control-flow of the business process.

*Control-Flow Footprint Update and Sub-Footprint Configurations.* The *Control-flow Footprint Update* processes the global, standardized events and updates the dynamic control-flow footprint. In addition it provides the possibility to create dynamic control-flow sub-footprints as specified in the *Sub-Footprint Configurations* in order to allow for set of activities to specifically discover their *eventually follows, direct neighbour*, etc. relationships (see 4.2: Control-flow Footprint).

### 4.5 Footprint Interpretation

The *Footprint Interpretation* is the agent that translates the state information in form of the dynamic footprint into state information that conforms to the business process notation (see Figure 2). As opposed to the run-time event processing agent it has less restrictive constraints with regards to run-time as it is only executed on demand, after a specified amount of time passed or after a specified number of events or traces occurred. However, since this framework was designed to enable real-time or near real-time analysis the run-time should not increase exponentially in relation to the number of activities involved.

Another interesting challenge we had to overcome during development was that for some footprints multiple interpretations existed, i.e. two different control-flows can produce the exact same traces. In these cases it could happen that the *state of the business process* was alternating between these options which is an undesired behaviour. To prevent this from happening in the SDPD framework, business knowledge was applied to assign priorities to interpretations that allowed selecting the interpretation with the highest priority in favour of equally possible interpretations.

*Control-Flow Interpretation* with the given run-time constraints is not a trivial matter. For this purpose a new algorithm has been developed that follows a top-down approach and operates in such a way that different possibilities, e.g. split, loop, decision, sequence, etc., compete with each other for the best fitting solution. If a parallel split or loop was detected in the footprint the creation of a sub-footprint is requested. After information for these sub-footprints could be obtained, the process can be completely extracted from the given footprints.

## 5 Conclusion

In this paper we introduced the concept of Scalable Dynamic Process Discovery as an adapted process discovery application of Event-driven Business Process Management. Scalable Dynamic Process Discovery describes the method of monitoring one or more BPMSs in order to provide at any point in time a reasonably accurate representation of the current state of the processes deployed in the systems with regards to their control-flow, resource, and performance perspectives as well as the state of currently open traces. One important feature of this concept is that it is independent from any additional model input, i.e. the important aspects of a process like control-flow, performance, etc. are mined during runtime and are, as defined in [14], part of the "state" of the process. Additionally, we presented a framework for this concept along with the description of involved agents and models. These descriptions include additional requirements and specifications, a few important implementation details and findings, and references to similar and existing work associated with occurring challenges. The presented framework is driven by the requirements of real life industrial use cases provided by business partners within the EU funded project TIMBUS.

During the evaluation in the context of the use-cases it became apparent that this concept still has a small number of limitations which are considered to be future work: (1) Changes in the state of the business process were usually detected almost immediately but it took a longer time until the new state of the system was reflected appropriately in the extracted business process model. This behaviour originates from the fact that the footprint and the interpreted business process are in a sort of intermediate state for a while until the influence of the old version of the business process has disappeared. (2) At the moment the framework only captures information of the current state of the business process. If the SDPD framework would also keep and store previous states then that would allow for reasoning not only on the basis of the current state but also on the history of the business process. That would positively affect, for instance, prediction results or evolution analyses. (3) The generalization requirement in the *Event Hub* component and the *Global, Standardized Events* model has been addressed in an arguably "naive" way only: For instance it has not been discussed how differences in some of the dimensions of the event formats can be overcome to transform into a unified event model. Examples of these differences are event granularity (activity level events vs. process level events) or perspective support (whether resource, trace or other information provided or not). (4) More meaningful results could be achieved if the data perspective provided by events would not be ignored but would instead be generalized and thus become a part of the *dynamic state* of the business process as well.

**Acknowledgement.** Project partially funded by the European Commission under the 7th Framework Programme for research and technological development and demonstration activities under grant agreement 269940, TIMBUS project (http://timbusproject.net/).

# References

1. von Ammon, R., Ertlmaier, T., Etzion, O., Kofman, A., Paulus, T.: Integrating Complex Events for Collaborating and Dynamically Changing Business Processes. In: Dan, A., Gittler, F., Toumani, F. (eds.) ICSOC/ServiceWave 2009. LNCS, vol. 6275, pp. 370–384. Springer, Heidelberg (2010)
2. von Ammon, R.: Event-Driven Business Process Management. In: Proceedings of Encyclopedia of Database Systems, pp. 1068–1071. Springer US (2009)
3. Eckert, M.: Complex Event Processing with XChange EQ: Language Design, Formal Semantics, and Incremental Evaluation for Querying Events (2008)
4. Friedenstab, J.-P., Janiesch, C., Matzner, M., Müller, O.: Extending BPMN for Business Activity Monitoring. In: Proceedings of 45th Hawaii International International Conference on Systems Science, pp. 4158–4167. IEEE (2012)
5. Günther, C.W., Verbeek, E.: XES - Standard Definition (2012), http://www.xes-standard.org/_media/xes/xesstandarddefinition-1.4.pdf (accessed January 25, 2014)
6. Intalio. BPMS designer, http://www.intalio.com/products/bpms/overview/ (accessed January 25, 2014)
7. Janiesch, et al.: Slipstream: Architecture Options for Real-time Process Analytics. In: Chu, W., et al. (eds.) Proceedings of the ACM Symposium on Applied Computing (2011)
8. Ko, R.K.L.: A computer scientist's introductory guide to business process management (BPM). ACM Crossroads Journal (2009)
9. Ko, R.K.L., Lee, S.S.G., Lee, E.W.: Business Process Management (BPM) Standards: a Survey. BPM Journal 15(5), 744–791 (2009)
10. Luckham, D.: The Power of Events: An Introduction to Complex Event Processing in Distributed Enterprise Systems. Addison-Wesley Professional, Reading (2002)
11. OASIS: Web Services Business Process Execution Language Version 2.0. (2007), http://docs.oasis-open.org/wsbpel/2.0/wsbpel-v2.0.pdf
12. Object Management Group Inc.: Business Process Model and Notation (BPMN) Specification 2.0 (2011), http://www.omg.org/spec/BPMN/2.0/PDF
13. Redlich, D., Gilani, W.: Event-Driven Process-Centric Performance Prediction via Simulation. In: Daniel, F., Barkaoui, K., Dustdar, S. (eds.) BPM 2011 Workshops, Part I. LNBIP, vol. 99, pp. 473–478. Springer, Heidelberg (2012)
14. Redlich, D., Blair, G., Rashid, A., Molka, T., Gilani, W.: Research Challenges for Business Process Models at Run-time. LNCS State-of-the-Art Survey Volume on Models@run.time (2014) (not published yet)
15. Scheer, I.D.S.: ARIS (Architecture of integrated Information Systems) (1992)
16. van der Aalst, W., Weijters, A., Maruster, L.: Workflow Mining: Discovering Process Models from Event Logs. IEEE Transactions on Knowledge and Data Engineering 16(9), 1128–1142 (2004)
17. van der Aalst, W.M.P., ter Hofstede, A.H.M., Weske, M.: Business Process Management: A Survey. In: van der Aalst, W.M.P., ter Hofstede, A.H.M., Weske, M. (eds.) BPM 2003. LNCS, vol. 2678, pp. 1–12. Springer, Heidelberg (2003)
18. van der Aalst, W., et al.: Process Mining Manifesto. In: Daniel, F., Barkaoui, K., Dustdar, S. (eds.) BPM 2011 Workshops, Part I. LNBIP, vol. 99, pp. 169–194. Springer, Heidelberg (2012)

19. van der Aalst, W., Ter Hofstede, A.: YAWL: Yet Another Workflow Language (2003)
20. van der Aalst, W.: Process Mining - Discovery, Conformance and Enhancement of Business Processes. Springer (2011)
21. Weijters, A., van der Aalst, W., Alves de Medeiros, A.: Process Mining with the Heuristics Miner-algorithm. BETA Working Paper Series, WP 166, Eindhoven University of Technology (2006)
22. Woods, D., Word, J.: SAP Netweaver for Dummies. Wiley, Hoboken (2004)
23. zur Muehlen, M., Swenson, K.D.: BPAF: A Standard for the Interchange of Process Analytics Data. In: Muehlen, M.z., Su, J. (eds.) BPM 2010 Workshops. LNBIP, vol. 66, pp. 170–181. Springer, Heidelberg (2011)

# From Business Process Models to Use Case Models: A Systematic Approach

Estrela Ferreira Cruz[1,2], Ricardo J. Machado[2], and Maribel Yasmina Santos[2,*]

[1] Instituto Politécnico de Viana do Castelo, Portugal
estrela.cruz@estg.ipvc.pt
[2] Centro ALGORITMI, Escola de Engenharia,
Universidade do Minho, Guimarães, Portugal
{rmac,maribel}@dsi.uminho.pt

**Abstract.** One of the most difficult, and crucial, activities in software development is the identification of system functional requirements. A popular way to capture and describe those requirements is through UML use case models. A business process model identifies the activities, resources and data involved in the creation of a product or service, having lots of useful information for developing a supporting software system. During system analysis, most of this information must be incorporated into use case descriptions. This paper proposes an approach to support the construction of use case models based on business process models. The proposed approach obtains a complete use case model, including the identification of actors, use cases and the corresponding descriptions, which are created from a set of predefined natural language sentences mapped from BPMN model elements.

**Keywords:** Business Process Modeling, BPMN, Use Case Model, UML.

## 1 Introduction

Markets' globalization and the constant increase of competition between companies demand constant changes in organizations in order to adapt themselves to new circumstances and to implement new strategies. Organizations need to have a clear notion of their internal processes in order to increase their efficiency and the quality of their products or services, increasing the benefits for their stakeholders. For this reason, many organizations adopt a business process management (BPM) approach. BPM includes methods, techniques, and tools to support the design, enactment, management, and analysis of operational business processes [1].A business process is a set of interrelated activities that are executed by one, or several, organizations working together to achieve a common business purpose [2]. Among the various existing modeling languages, we opted for the Business Process Model and Notation (BPMN), currently in version 2.0 [3], because it is a widespread OMG standard that is actually used both in academia and in organizations.

---

[*] This work has been supported by FCT - Fundação para a Ciência e Tecnologia within the Project Scope: PEst-OE/EEI/UI0319/2014.

If on one hand the business process management and modeling are increasing their relevance, on the other hand the software development teams still have serious difficulties in performing elicitation and defining the applications requirements [4]. In fact, one of the main software quality objectives is to assure that a software product meets the business needs [4]. For that, the software product requirements need to be aligned with the business needs, both in terms of business processes and in terms of the informational entities that those processes deal with. This drives us to the question: "Can the existing model information about business processes be used as a basis for modeling the software applications that support that business?"

Information systems researchers and professionals have recognized that understanding a business process is the key to identify the user needs of the software that supports it [5,6]. However, the tasks of business process analysis and software development are managed by different groups of people and commonly use different languages.

Requirement elicitation is, indeed, a key step in the software development process. Use case models aim to capture and describe the functional requirements of a system [7]. Dietz says that the use cases strong point is that once they are identified, the development of the software application goes well [8]. The weak point is the identification of use cases themselves. Shishkov *et al.* states that deriving use case models from business analysis models would be useful, since both reflect behavior within business/software systems [6].

A use case model is a set of use case diagrams and the corresponding use case descriptions [9]. The use case diagrams enable to perceive the need of describing the system behavior in response to messages received from outside the system (i.e., from its actors) [10].

In this paper, we present an approach to obtain a complete use case model based on a business process model. All information existing in a BPMN model that cannot be represented as an actor or as a use case will be depicted as textual use case description. Use case descriptions are, commonly, specified in Natural Language (NL) [11,12]. As Fantechi *et al.* say NL is easy to understand but, at the same time, could be ambiguous, redundant and with omissions [11]. However, the generated descriptions are a set of controlled sentences previously defined in NL.

The remainder of this paper is structured as follows. In the next section, BPMN and basic concepts of use case models are introduced and some related work is presented. Section 3 describes our approach for use case model creation and presents its application to an example. Finally, conclusions and some remarks to future work are presented.

## 2 Background

### 2.1 The BPMN Language

Business process management focus its attention on designing and documenting business processes, in order to describe which activities are performed and the dependencies between them [13]. The BPMN basic process models can be grouped into two types of processes [3]:

- **Private Business Processes** - A private process is a process internal to a specific organization. Each private process is represented within a Pool. The process flow must be in one pool and should never cross the boundaries of that Pool. The interaction between distinct private Business Processes can be represented by incoming and outgoing messages.
- **Public Processes** - A public process represents the interactions between a private Business Process and other Processes or Participants. Only activities that are used to communicate with the other participants must be included in the public process.

The BPMN's diagrams use a set of graphical objects that can be grouped into five basic categories [3]:

- **Flow Objects** - are the main graphical elements to define the behavior of a Business Process. There are three kinds of Flow Objects: Events, Activities and Gateways.
- **Data** - represent the data involved in the process. Data that flows through a process is represented by *data objects*. Persistent data can be represented by *data stores*. Data objects and data stores are exclusively used in private process diagrams [3].
- **Connecting Objects** - model the connection between the several process elements. There are four types of connecting objects: Sequence Flows, Message Flows, Associations and Data Associations.
- **Swimlanes** - represent the participants in the process. A participant is a person, or something, involved in the process. Participants in the process can be grouped into pools or, more particularly, in Lanes. A pool can be divided into several Lanes, for example, to represent the different departments of an organization involved in the process.
- **Artifacts** - are used to provide additional information to the process, such as a note ("Text Annotation").

During a process execution, resources and/or data are consumed and produced. The transmission of the data created or used during a process execution can be represented by *Messages* or *Data Associations*.

The following subsection addresses use case models.

## 2.2 Use Case Model

Booch *et al.* say that use case models, when defined by Ivar Jacobson, aimed to describe the behavior of the system from the users point of view [14]. So, it is expected that a use case model specifies what a system is supposed to do [15]. In [15] a use case is defined as a *behavioral classifier* that represents a declaration of a set of offered behaviors. Each use case specifies some behavior, possibly including variants, which the subject can perform in collaboration with one or more actors.

A use case model should identify the system boundaries (depicted as a rectangle) and the actors, which are represented by a "stickman" icon outside the system boundaries [7,15]. An actor is someone or something that interacts with the system [15]. So, an actor is always related to one or more use cases. A use case is graphically represented by an ellipse and contains a brief description of the action [9]. A use case diagram is composed by actors and use cases. Each use case shall have an associated description. There are some alternatives that can be used to describe a use case, like informal text, numbered steps, pseudo-code, among others [12]. Cockburn proposes a basic use case descriptions template that includes the use case name, actors, scope, context, pre-conditions, primary success scenario, alternate scenarios, amongst others [12].

### 2.3 Existing Approaches

It is recognized that the software that supports the business must be aligned with the business processes [16]. Therefore, it is natural to try an approximation between business process modeling and software modeling. Requirements elicitation is usually the first phase on a software development process. Several authors already propose approaches to derive use cases from business process models. Some of the existing approaches are presented next.

Dijkman and Joosten propose an approach that maps a business process model (modeled using the UML Activity Diagram) into use case diagrams [17]. They also proposed an algorithm to derive a use case diagram from a business process modeled as activity diagrams [18]. To do so, Dijkman and Joosten start by defining the activity diagram and the use case diagram meta-models. Then, the authors establish a relation between the "role" from the activity diagram and the "actor" in a use case diagram and a "step" (a sequence of tasks) from the activity diagram originates a "use case" in a use case diagram [18].

Rodriguez *et al.* propose a systematic approach to derive a use case diagram from a UML activity diagram [19] and another to derive a use case diagram from a BPMN model [20]. In the latter approach, the transformation is guided by a set of QVT (Query View Transform) rules and checklists. In a summarized way, in Rodriguez *et al.* approach, a participant is mapped to an actor in the use case diagram; an activity in the BPMN model gives origin to a use case.

All surveyed existing approaches obtain a use case diagram based on a business process model, but no one presents a proposal for obtaining the use cases description. Nevertheless, the use cases descriptions are one of the most important components of the use case model [12,21]. Moreover, without descriptions most information presented in a business process model will be lost when generating the use case diagram from a business process model.

Cockburn emphasizes the use case descriptions. In Cockburn's opinion the use case writers should spend their time and effort on use case descriptions [12]. The use case descriptions can specify all information needed. But, how should the use cases be written? Cockburn advises the use case writers to use sentences with a simple structure, which should be "easy to read and follow" [12] and describes a semi-formal structure to use cases description.

The CREWS (Co-operative Requirements Engineering With Scenarios) team proposes two sets of guidelines to be used on use case descriptions: six guidelines related to style and eight related to content [22]. Karl Cox also presents a set of structure guidelines for use case descriptions [23]. More exactly he proposes the *CP Use Case Writing Rules*, a small set of guidelines derived from the 7 C's (Coverage, Cogent, Coherence of logic, Consistent abstraction, Consistent Structure, Consistent Grammar, Consideration of alternatives) [23].

Comparing CREWS and CP guidelines, the CP guidelines number is smaller and intends to be easier to apply than CREW guidelines [24]. Both provide improvements on use case descriptions quality [24] and subsequently improve the understanding between stakeholders.

The next section describes our approach to obtaining the use case model from a business process model.

## 3 The Proposed Approach

Graphically a use case diagram is very simple because it only involves actors and use cases (stickman's and ellipses with a brief description). A BPMN process diagram is graphically more complex because it involves lots of graphical elements (activities, events, gateways, data objects, pools, etc.). However a use case model can represent as much information as a BPMN model, but most of the information must be embodied in use case descriptions. So, the approach presented here is specially focused on use case descriptions for which we present a template.

The approach is divided in two main parts. First we present a set of rules to obtain a use case diagram from a BPMN model. Then we address the rules to derive the description of the uses cases previously identified.

### 3.1 Use Case Diagram Generation

The presented approach is based on the private business process, where messages exchanged with other participants, or business partners, shall be represented. The proposed approach is based on the following considerations:

- The information about the participants in the process is relevant to the process, so all participants involved in messages exchange must be represented.
- An activity represents some work performed within a business process. An activity may be atomic, usually represented as a task, or non-atomic, represented as a sub-process. To avoid information loss during the application of the proposed approach, the sub-processes must be expanded.
- A manual task is a task performed without any information technology involvement [25]. Nevertheless, the information about the task execution, like start and ending time or amount of resources produced and consumed, can be useful to the process monitoring to support and evaluate future decisions or improvements.

We agree with Rodriguez *et al.* on mapping a participant to an actor and one activity to a use case [20]. Accordingly, the rules to generate the use case diagram are explained below:

- R1: A role played by a participant (represented by a lane or a pool) must be represented by an actor in the use case diagram. The actor name is the participant name.
- R2: A lane can be the sub-division of a pool or a sub-division of another lane. These subdivisions form the actors' hierarchy:
  - If the lane is a sub-division of a pool then the actor that represents the lane is a specialization of the actor that represents the pool;
  - If the lane is a sub-division of another lane then the actor that represents the internal lane is a specialization of the actor that represents that lane.
- R4: Each activity will be represented as a use case in the use case diagram. The use case name (brief description of the action) is the activity name.
- R5: An actor that represents a pool (or a lane) is related with all use cases representing the activities that belong to the pool (or lane).
- R6: The actor that represents the participant that sends (or receives) a message to an activity is related to the use case that represents that activity.

Next subsection applies the described rules to the Nobel Prize example.

### 3.2 Nobel Prize Example

The diagram shown in Figure 1, represents the Nobel Prize BPMN Process Diagram. The presented BPMN model comprises ten activities, consequently (by rule R4 above) there will be ten use cases on the generated use case diagram. Four pools are involved in the process: *Nobel Committee*, *Nominators*, *Expert* and *Nobel Assembly*. By R1 the obtained use case diagram will have four actors with the corresponding names. The obtained Nobel Prize use case diagram is shown in Figure 2.

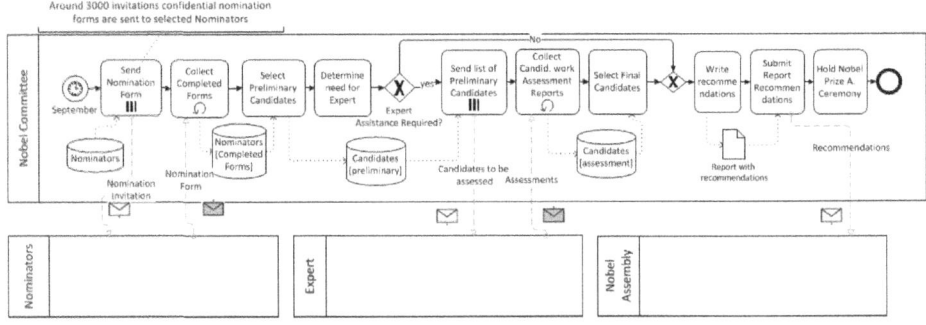

**Fig. 1.** The Nobel Prize Process Diagram (adapted from [26])

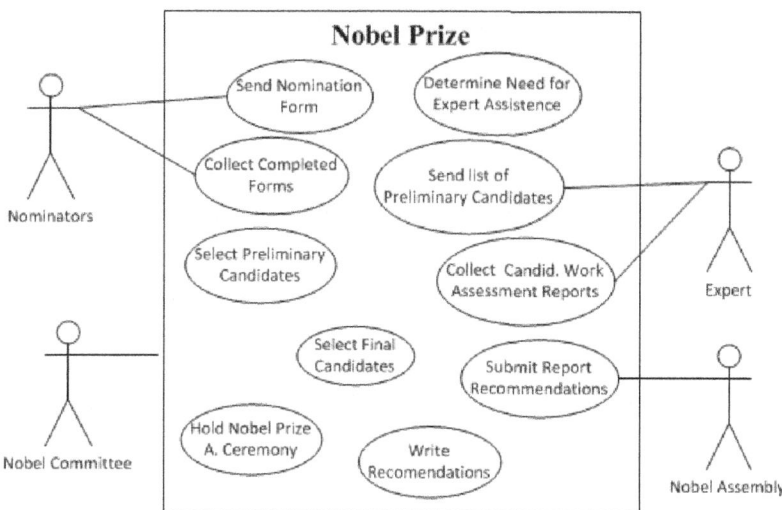

**Fig. 2.** The Nobel Prize Use Case Diagram

As can be seen in Figure 1, all activities are performed by *Nobel Committee* participant, so, by R5, all use cases are related with *Nobel Committee* actor. The *Nominators* participant sends a message to *Send Nomination Form* activity, so, by R6, the *Nominators* actor is related with the *Send Nomination Form* use case. The *Collect Completed Forms* activity receives a message from the *Nominators* pool, so, by R6, the *Nominators* actor is related with the *Collect Completed Forms* use case. The explanation for the other relationships is similar.

### 3.3 Getting Use Case Descriptions

This subsection addresses the generation of use case descriptions from a private business process model. We define a template to represent a use case description based on a simplification of the template presented by Cockburn in [12]. The proposed template is composed by six fields, which are named and described in Table 1.

Cockburn says that a real big and complex system can be modeled with only seven use cases [12]. This yields very complex use cases with several alternative scenarios. Our approach, by transforming each BPMN activity into a different use case, yields much simpler use cases, each with a single scenario. For that reason the proposed template only attend to one (main) scenario. Pre-conditions, triggers and post-conditions enable the representation of the process flow in the use case model.

The main elements involved in a process are participants (pool and lanes), activities, gateways, events, messages, data objects, data stores and artifacts [3]. These elements are connected by connecting objects (sequence flow, message flow, associations and data associations). The approach being presented intends

Table 1. The template for describing use cases

| | |
|---|---|
| **Use Case name** | The use case name identifies the goal as a short active verb phrase. |
| **Actors** | List of actors involved in the use case |
| **Pre-Conditions** | Conditions that must hold or represent things that happened before the use case starts. |
| **Post-Conditions** | Conditions that must hold at the conclusion of the use case. |
| **Trigger** | Event that starts the use case. |
| **Scenario** | Sequence of interactions describing what the system must do to move the process forward. |

to transform business process elements, and their associated information, in a controlled set of sentences in NL, following the CREWS guidelines.

The activity name is the use case name in the use case template. The related pools or lanes represent the actors related with the use case in the use case template, as described in sub-section 3.1.

Focusing our attention on a use case, all incoming connections and outgoing message flows, data associations, and sequence flows to events of the corresponding activity must be reflected in the use case descriptions, fulfilling the use case template previously defined.

Sequence flows outgoing an activity to a gateway or to another activity do not create a sentence in the source activity description because these connections already create sentences in the activity that receives the sequence flow.

Each connecting object makes a connection between a source (sourceRef) and a target (targetRef). Different connecting objects connect different elements. The next sub-sections describe how incoming and outgoing connections of an activity are represented in the corresponding use case template.

**Data Associations.** Data associations are used to move data between data objects (or data stores) and activities [3]. The data (physical document or information) that are created, manipulated, and used during the execution of a process are represented as data objects (or data object references) or as data stores (or data store references). A data object reference is a way to reuse data objects in the same diagram [3]. The same happens with the data store reference.

The sentences generated by data associations and associated data objects, or data stores, are represented in Table 2. The sentences will be appended to the scenario of the use case description of the use case that represents the activity.

**Association.** An association is used to link text annotations and other artifacts with other BPMN graphical elements [3]. When an association links a text annotation with an activity, the text is transcribed to the scenario of the use case that

**Table 2.** The use case sentences originated by Data Associations

| Data | Graphical representation | Originated sentence in use case scenario. |
|---|---|---|
| **Data Object** as data association source | | Receives <data object name>. |
| **Data Object** as data association target | | Sends <data object name>. |
| **Data Input** | | Receives <data object name>. |
| **Data Input Collection** (Input set) | | Receives a collection of <data object name>. |
| **Data Output** | | Sends <data object name>. |
| **Data Output Collection** (Output set) | | Sends a collection of <data object name>. |
| **Data Store** as data association source | | Reads information from <data store name> |
| **Data Store** as data association target | | Writes information on <data store name> |

represents the activity. The text remains the same. When an association links a text annotation to a gateway, or to a sequence flow, the text is transcribed to the scenario of the use case that represents the target activity.

**Message Flow.** A message flow connects two pools representing the message exchange between the two participants [3]. A message represents the content of a communication between two Participants [3]. A Message is graphically represented as an envelope as we saw in Figure 1. The sentences originated by a message flow are described next as two different rules (MR1 and MR2).

- MR1: When an activity receives a message (message input), the use case that represents the activity will have the following sentence in its use case scenario: **Receives <message name> [with <messageRef>] from < participant name>**.
- MR2: When an activity sends a message (message output), the use case that represents the activity will have the following setence in its use case scenario: **Sends <message name> [with <messageRef>] to <participant name>**.

*MessageRef* defines the message that is passed via message flow. It can be any kind of information exchanged between different pools (an email, a phone call, a document, etc.).

**Sequence Flow.** A sequence flow is used to show the order that activities are performed in a process [3]. A sequence flow can connect activities, events and gateways [3]. When a sequence flow connects two activities, it originates the next sentence as pre-condition in the use case that represents the target activity: **The <source activity name> has been completed.**

Everything that occurs between two activities must be registered in the target activity description. Involved gateways and events are treated in the next sub-sections.

**Sequence Flow and Gateways.** Gateways are used to control how the process flows, by diverging (splitting gateways) and converging (merging gateways) sequence flows. Splitting gateways have one incoming sequence flow and two or more outgoing sequence flows. Merging gateways have two or more incoming sequence flows and one outgoing sequence flow [3], as we can see in Table 3.

The gateway's outgoing sequence flows may have a *Condition* that allows to select alternative paths. Each outgoing sequence flow originates a sentence represented as a pre-condition in the use case description of the sequence flow target activity. The generated sentences are represented in Table 3.

**Table 3.** The use case pre-condition originated by gateways

| Gateway | Graphical representation | Originated **Pre-condition.** |
|---|---|---|
| **Exclusive Decision** | | The <gateway condition> is <sequence flow condition>. |
| **Parallel splitting** | | The <source name> has been completed. |
| **Inclusive Splitting** | | The <sequence flow condition> is true. |
| **Complex Splitting** | | The <sequence flow condition> is true. |
| **Exclusive merging** | | The <source name> [exclusive or <source2 name>] has been completed. |
| **Parallel join** | | The <source name> [and < source2 name>] has been completed. |
| **Inclusive merging** | | The <source name> [ or <source2 name>] has been completed. |
| **Complex merging** | | The <source name> [or <source2 name>] has been completed. |

**Sequence Flow and Events.** An event is something that happens during the course of a process and that affects the process's flow [3]. These events usually have a cause or produce an impact [3]. In BPMN 2.0 there is a large number of event types, so we present a general overview of the generic sentences originated in the use case template by the different events categories (see Table 4). Each category has its own table to address differences that can exist between sentences generated by the events of the same category. Due to lack of space only the sentences generated by catching events are presented here (Table 5).

Table 4. Generic sentences originated by events

| Event type category | Generic sentence originated in use case template |
|---|---|
| Start | **Trigger:** The <event name - event definition> occured. |
| Intermediate Catching | **Trigger:** The <event name - event definition> is received. |
| Intermediate Boundary Interrupting | **Scenario:** If the <event name - event definition> occurs, the <activity name> is interrupted. |
| Intermediate Boundary Non-Interrupting | **Scenario:** The <event name - event definition> occurred. |
| Intermediate Throwing | **Post-condition:** The <event name - event definition> is created. |
| End | **Post-condition:** The <event name - event definition> is created. The process ends. |

The events affect the sequence or the timing of the process's activities. There are three types of events: Start, Intermediate and End. Start events indicate where a process (or a sub-process) will start. End events indicate where a path of a process will end. Intermediate events indicate where something happens somewhere between the start and end of a process [3].

Some events are prepared to catch triggers. These events are classified as catching events. Events that throw a result are classified as throwing events. [3]. All start events and some intermediate events are catching events[3]. The sentence originated by a catching event is included as a trigger in the description of the use case that represents the activity that is started by the event. Catching events are represented as triggers because this events cause the start of the activity.

All end events and some intermediate events are throwing events [3]. The sentences originated by the throwing events are included as a post-condition in the description of the use case that represents the activity that throws the event. Throwing events are represented as a post-condition because the event is a consequence (or a result) of the activity execution.

**Table 5.** The sentences originated by catching events

| Catching Event | Originated sentence in use case **trigger**. |
|---|---|
| None | The event <event definiton> occurs. |
| Message | The message <event definition> arrives from <source>. |
| Timer | The time-date <event definition> is reached. |
| Conditional | The condition <expression> become true. |
| Signal | The signal <event definition> arrives. |
| Multiple | The <event definition> [or <event definition>] occurs. |
| Parallel Multiple | The <event definition> [and <event definition>] occurs. |

Some events can also be classified as interrupting or non-interrupting events. Interrupting events stop its containing process whenever the event occurs. When Non-Interrupting events occur its containing process is not interrupted [3].

An event can be thrown by an activity and caught by another. In this case the event originates a sentence in the post-condition of the use case representing the activity that throws the event and another sentence in the trigger of the use case representing the activity that catches the event.

In the next subsection the defined approach is applied to the Nobel Prize example.

### 3.4 Nobel Prize Example

For reasons of space, we cannot show the complete example here. So, we select the use cases that cover a greater number of application cases.

As we can see in Figure 1, the *Send Nomination Form* activity has four incoming connections: a sequence flow from an event, giving origin to a sentence in use case trigger (Table 5), an incoming message flow, a data association and an association, each one generating a sentence in use case scenario. The corresponding use case descriptions are presented in Table 6.

The *Send List of Preliminary Candidates* activity has two incoming connections: a sequence flow from a gateway, giving origin to a pre-condition (Table 3) and a data association giving origin to a sentence in use case scenario. The activity also has an outgoing message flow to *Expert* participant generating a sentence in use case scenario (Table 2). The corresponding use case descriptions are presented in Table 7.

The *Write recommendations* activity has an incoming sequence flow from a gateway, giving origin to a pre-condition (Table 3) and an outgoing data

**Table 6.** Send Nomination Form use case description

| Use Case name | Send Nomination Form. |
|---|---|
| Actors | Nobel Committee, Nominator |
| Trigger | The time-date September is reached. |
| Scenario | Around 3000 invitations confidential nomination forms are sent to selected Nominators. Reads information from Nominators. Sends the Nomination Invitation to Nominator. |

**Table 7.** Send List of Preliminary Candidates use case description

| Use Case name | Send List of Preliminary Candidates. |
|---|---|
| Actors | Nobel Committee, Expert |
| Pre-condition | The Expert Assistance Required? is Yes. |
| Scenario | Reads information from Preliminary Candidates. Sends the List of Candidates to be Assessed to Expert. |

**Table 8.** Write Recommendations use case description

| Use Case name | Write Recommendations. |
|---|---|
| Actors | Nobel Committee |
| Pre-condition | The Expert Assistance Required? is No or Select Final Candidates has been completed. |
| Scenario | Sends The Report with Recommendations. |

association giving origin to a sentence in the use case scenario. The corresponding use case descriptions are presented in Table 8.

## 4 Conclusions and Future Work

This paper presents an approach to generate a use case model, including descriptions, from a private BPMN process diagram. The approach starts by presenting a set of rules to generate the use case diagram in which each activity in the BPMN model gives origin to a use case and a participant gives origin to an actor in use case model. To identify the use cases description a set of structured sentences are created in NL. Each sentence represents an incoming or outgoing connection from the use case corresponding activity.

BPMN has originally been design to be a language easy to understand by all stakeholders involved [27,3], nevertheless with the increase in number of its graphical elements, in its most recent version (BPMN2.0), the language has become more complex and consequently difficult to understand. The approach presented herein helps understanding BPMN models as it translates a model to NL, promoting the understanding between the involved stakeholders.

The BPMN2.0 allows business process models to be highly detailed. This is good news if one intends to use BPMN models as a basis to the development of the software that supports the business. The presented approach benefits from a detailed business process model, as greater business process detail yields a more complete use case model.

Generating a complete use case model from a business process model allows us to use existing methods, techniques and tools to generate other software models from use case models. One of those methods is the 4SRS (4-Step Rule Set), which generates a logical architecture and corresponding class diagrams from user requirements, represented as use cases [28]. The presented approach enables traceability between business processes and the corresponding elements in software models.

Typically, in a real situation, a software product does not support only one business process, but rather a set of processes. So, in order to generate a complete use case model for the development of such software product, we intend to extend the approach presented herein to generate a use case model representing the set of processes that comprise a business.

When sub-processes are involved, this approach demands that they are fully expanded, losing some structuring information. As future work, we intend to treat the sub-processes by refining the use cases in different detail levels.

# References

1. van der Aalst, W.M.P.: Business process management demystified: A tutorial on models, systems and standards for workflow management. In: Desel, J., Reisig, W., Rozenberg, G. (eds.) ACPN 2003. LNCS, vol. 3098, pp. 1–65. Springer, Heidelberg (2004)
2. Ko, R.K.L.: A computer scientist's introductory guide to business process management (bpm). Crossroads 15, 11–18 (2009)
3. OMG, Business process model and notation (BPMN), version 2.0. Tech. rep., Object Management Group (2011)
4. Jalote, P.: A concise Introduction to Software Engineering. Springer (2008)
5. Mili, H., Jaoude, G.B., Lefebvre, É., Tremblay, G., Petrenko, A.: Business process modeling languages: Sorting through the alphabet soup. In: OOF 22 NO. IST-FP6-508794 (PROTOCURE II) (September 2003)
6. Shishkov, B., Xie, Z., Liu, K., Dietz, J.L.: Using norm analysis to derive use cases from business processes. In: Proc. 5th Workshop on Organiz. Semiotics (2002)
7. Hull, E., Jackson, K., Dick, J.: Requirements Engineering. Springer (2011)

8. Dietz, J.L.G.: Deriving use cases from business process models. In: Song, I.-Y., Liddle, S.W., Ling, T.-W., Scheuermann, P. (eds.) ER 2003. LNCS, vol. 2813, pp. 131–143. Springer, Heidelberg (2003)
9. Bittner, K., Spence, I.: Applying use cases: a practical guide. P. Ed. inc. (2003)
10. Roussev, B.: Generating OCL specifications and class diagrams from use cases: a newtonian approach. In: Proceedings of the 36th Annual Hawaii International Conference on System Sciences, p. 10 (January 2003)
11. Fantechi, A., Gnesi, S., Lami, G., Maccari, A.: Applications of linguistic techniques for use case analysis. Req. Eng. 8(3), 161–170 (2003)
12. Cockburn, A.: Writing Effective Use Cases. Addison Wesley (2001)
13. Meyer, A.: Data in business process modeling. In: Proceedings of the 5th PhD Retreat of the HPI Research School on Service-oriented Systems Engineering (2010)
14. Booch, G., Rumbaugh, J., Jacobson, I.: The Unified Modeling Language User Guide. Addison Wesley (1998)
15. OMG, Unified modeling language (OMG UML), version 2.5. Tech. Rep., Object Management Group (2012)
16. Giaglis, G.M.: A taxonomy of business process modeling and information systems modeling techniques. International Journal of Flexible Manufacturing Systems 13, 209–228 (2001)
17. Dijkman, R.M., Joosten, S.M.: Deriving use case diagrams from business process models. Tech. rep., CTIT Tech. Rep., Enschede, The Netherlands (2002)
18. Dijkman, R.M., Joosten, S.M.: An algorithm to derive use cases from business processes. In: 6th Int. Conf. on Software Engineering and Applications (2002)
19. Rodríguez, A., Fernández-Medina, E., Piattini, M.: Towards obtaining analysis-level class and use case diagrams from business process models. In: Song, I.-Y., et al. (eds.) ER 2008 Workshops. LNCS, vol. 5232, pp. 103–112. Springer, Heidelberg (2008)
20. Rodríguez, A., Fernández-Medina, E., Piattini, M.: Towards CIM to PIM transformation: From secure business processes defined in BPMN to use-cases. In: Alonso, G., Dadam, P., Rosemann, M. (eds.) BPM 2007. LNCS, vol. 4714, pp. 408–415. Springer, Heidelberg (2007)
21. Bittner, K., Spence, I.: Use Case Modeling. Pearson Education Inc. (2003)
22. Rolland, C., Achour, C.B.: Guiding the construction of textual use case specifications. Data & Knowledge Engineering 25, 125–160 (1998)
23. Cox, K.: Heuristics for use case descriptions. Thesis (PhD) (November 2002)
24. Phalp, K., Vincent, J., Cox, K.: Improving the quality of use case descriptions: empirical assessment of writing guidelines. Software Quality Journal 15(4), 383–399 (2007)
25. Allweyer, T.: BPMN 2.0 - Introduction to the standard for business process Modeling. Books on Demand GmbH, Norderstedt (2010)
26. OMG, BPMN 2.0 by example. Tech. Rep., Object Management Group (2010)
27. Magnani, M., Montesi, D.: BPDMN: A conservative extension of BPMN with enhanced data representation capabilities. In: CoRR (2009)
28. Santos, M.Y., Machado, R.J.: On the derivation of class diagrams from use cases and logical software architectures. In: 2010 Fifth ICSEA (2010)

# Approach for Semi-automatic Extraction of Business Vocabularies and Rules from Use Case Diagrams[*]

Tomas Skersys, Paulius Danenas, and Rimantas Butleris

Center of Information Systems Design Technologies,
Department of Information Systems, Kaunas University of Technology
Studentu str. 50-313a, Kaunas, Lithuania
{tomas.skersys,paulius.danenas,rimantas.butleris}@ktu.lt

**Abstract.** The main purpose of this paper is to explore the possibilities to extract well-structured business vocabularies and rules from the formalized requirements specifications expressed via use case diagrams; Object Management Group's (OMG) standards, namely *Semantics of Business Vocabularies and Business Rules* (SBVR) and *Unified Modeling Language* (UML), are used for this purpose. The paper concentrates on a semi-automatic extraction approach by proposing UML2SBVR mapping matrix, extraction algorithm and implementation prototype. An experiment and the evaluation of its results are discussed to prove the usability of the presented approach.

**Keywords:** UML Use Case Diagram, SBVR Business Vocabulary and Business Rules, model extraction, model-to-model transformation.

## 1 Introduction

Use Case Model (UCM) is an essential artifact in system's functional requirements analysis and specification. At the core of UCM is a Use Case Diagram (UCD), which provides visual representation of possible interactions of a system, specified by the functionality it provides, and actors, who use that system to achieve their goals. Next to Class and Activity Diagrams, the Use Case Diagram is arguably the most popular diagram of OMG's Unified Modeling Language (UML) [17]; therefore, it is no surprise that this diagram, and the model as a whole, is being extensively discussed and researched within information systems (IS), as well as enterprise engineering disciplines where UML is widely used.

In this paper, one particular aspect of UCD application is discussed, namely, extraction of business vocabularies and rules (BV&R) from use case diagrams using semi-automatic model-to-model (M2M) transformation approach (UCD → BV&R). The development of this approach is one of the tasks of the ongoing VEPSEM project*, which integrates OMG standards BPMN [12], SBVR [15] and UML into

---

[*] The work is supported by the project VP1-3.1-ŠMM-10-V-02-008 "Integration of Business Processes and Business Rules on the Base of Business Semantics (VEPSEM)" (2013-2015), which is funded by the European Social Fund (ESF).

one modeling approach. The project follows basic principles of OMG's Model Driven Architecture (MDA) [14], which underlines the integration of models and M2M transformations as core features of model-driven information systems development (ISD).

According to MDA, early stage of ISD resides at a business level – this is a domain of business people who tend to communicate their business knowledge using natural language and are quite cautious about formal graphical models. However, due to its nature, there is always a risk of ambiguity and miscommunication when using any natural language. To reduce such risk, business knowledge can be expressed using dedicated formal languages, which are based on natural language. "Semantics of Business Vocabularies and Business Rules" (SBVR) [15] is a novel standard enabling one to express business knowledge using controlled natural language, which would be unambiguous and understandable to business and IT people and also interpretable by computers (e.g. for M2M transformation purposes). Generally, SBVR specification is organized in two vocabularies: a business vocabulary (BV) consisting of noun and verb concepts, and business rules (BR), which can be structural or operational.

In model-driven ISD, business modeling is followed by system analysis stage. Here, we have a UCM, which is one of those models playing the role of a bridge connecting business model to system design model. UCM greatly contributes to a common understanding between business and IT people of what the business expects from the future information system functionality-wise. Again, essential part of this model is a graphical UCD. As far as business people are involved, there will always be a risk of those people reading and interpreting formal graphical models incorrectly. Therefore, we argue that graphical diagrams should be accompanied by formalized textual specifications, which interpret those diagrams in natural language-like format (e.g. SBVR). One of the ways to do it is to extract those textual specifications from diagrams themselves.

Moreover, ISD practice shows that it is a quite common case when systems development projects lack full scale business models, which would incorporate well-structured vocabularies of business concepts and rules as separate, manageable artifacts. In its turn, this leads to miscommunication among the project team members, inconsistency issues among the models at different stages of system development et cetera. The possibility to extract such vocabularies from UCD and use them later throughout the whole ISD life cycle would be very welcomed in such cases.

## 2 Related Work

At the moment of writing, we could not find any published research directly dealing with UCD → BV&R transformations. Also, there are only few papers presenting other kinds of transformations to obtain SBVR specifications from UML models or other knowledge representations, e.g. Cabot et al. [4] does that by transforming UML/OCL conceptual data models; in [10] and [11], unstructured natural language is used as a source of knowledge for the extraction of SBVR BV&R.

Following the systems development life cycle, a more common transformation is to the opposite direction, i.e. from business level SBVR specifications to models of system analysis and design. Though, it should be mentioned that research works dealing with such transformations do not have direct impact over our research – this is due to the fact that transformations involving SBVR specifications are, in general, not bidirectional ones. Nonetheless, these works give us some insights on the subject as such. Currently, there is only the research of Thakore and Upadhyay [19], which directly deals with BV&R → UCD transformation; the extraction of only the basic UCD elements is presented in that paper. Some works of this area of research are based on natural language patterns, which are semantically close to SBVR. Deeptimahanti and Sanyal [5], Kärkkäinen et al. [8], Georgiades and Aandreou [6] present approaches for the generation of UCD from such natural language specifications. In [6], the UCD transformation also includes various types of relationships to help generate more complete diagrams.

## 3    Choosing the Level of Automation of M2M Transformation

When undertaking any M2M (or M2Text, Text2M) transformation task, the first thing one must decide on is the automation level of the transformation, i.e. manual, semi-automatic or fully-automatic:

– Distinctive feature of *manual* M2M transformation is that it is done exclusively by the user, who might use his own empiric knowledge ("know-how") and/or formally defined transformation rules and algorithms to perform this task. Due to the lack of any automation, this kind of transformation is of no particular interest to us and will not be discussed any further.
– In case of *semi-automatic* M2M transformation, the transformation algorithm requires certain degree of user interaction to complete the task. Transformations of this kind are usually implemented as hard-coded applications.
– *Automatic* M2M transformations are capable to perform a fully-automated transformation of source models to target models without the involvement of a user in the transformation process. Automatic transformations may be implemented as hard-coded applications or by using dedicated transformation languages. Provided both source and target models have corresponding meta-models, the latter implementation technique is recommended.

As mentioned earlier, semi- and fully-automated M2M transformation approaches may be implemented using different implementation techniques. The most popular techniques are hard-coding and dedicated transformation languages-based (DTL) implementation, which involves the use of specialized M2M transformation engines.

*Hard-coding* technique does not use any intermediate technology to implement transformation. The transformation itself follows the programmed algorithm and is embedded into application. Due to its algorithmic nature, this technique might be considered as optimal solution for algorithm-based approaches involving user assistants (wizards). Wizard-assisted M2M transformation applications provide high level of flexibility and customization for the user. One of the main drawbacks of such implementation (and thus, the approach itself) lies in the management of changes at meta-models level – these changes must be implemented by altering the source code, which is time and effort consuming.

The latter drawback of hard-coding technique has much less impact on the *DTL* technique. DTL technique uses dedicated transformation languages to realize M2M transformations on meta-models level. This technique is recommended for the fully-automated M2M transformation approaches because DTL transformations do not provide the possibility of user interaction – in some cases, this might be considered as a drawback of this technique and fully-automatic approach as a whole. In general, approaches with high level of automation rely heavily on so called best modeling practices, which vary depending on a modeling language used (e.g. modeling business processes with UML and BPMN).

QVT (Query-View-Transform) [13] and ATL (ATLAS Transformation Language) [3] are arguably the most widely used dedicated transformation languages. QVT language is an OMG standard; its syntax is compatible with MOF 2.0. QVT defines three sublanguages for transforming models: declarative *Relations* and *Core* languages to specify transformations as a set of relations (object patterns) between model concepts, and imperative *Operational Mappings* language, which can be used as an extension to *Relations* language to specify additional constructs using imperative language elements (e.g. conditional statements, loops). Model querying and navigation is done using OCL language, which is also an OMG standard. Similarly to QVT, ATL language also provides both declarative and imperative constructs. Declarative constructs are expressed in a form of rules consisting of certain source and target patterns for source and target models' matching. ATL also supports rule inheritance and polymorphic rule reference [7].

As it was mentioned earlier, M2M transformation approaches, which are implemented using DTL, also require a supplementary transformation engine – in some situations, this could be considered as a drawback or even a deciding negative factor.

In VEPSEM project, both semi-automatic and automatic UCD → BV&R approaches are being researched. However, due to the limitations on paper size, only the semi-automatic approach with hard-code implementation will be presented further in this paper.

## 4  Approach for Semi-automatic Extraction of Business Vocabularies and Rules from Use Case Diagrams

### 4.1  The M2M Mapping Matrix

In order to develop M2M transformations (both hard-coded and DTL-based), a set of mappings between the elements of corresponding meta-models have to be identified. To specify the mappings among the relevant elements of UML meta-model (defining use case model) and SBVR meta-model, we follow a slightly improved approach as presented in [18]. The set of mappings is presented in a form of mapping matrix (Table 1). The meaning of markings $A_{\{i\}}$ and $M_{\{i\}}$ used in the matrix is as follows:

- "$A_{\{i\}}$" implies that a corresponding SBVR general concept, verb concept or rule can be *automatically* identified and extracted from one or more UCD elements;
- "$M_{\{i\}}$" implies *manual (semi-automatic)* identification and extraction of a corresponding SBVR general concept, verb concept or rule from one or more UCD elements;
- Each UCM concept is supplied with unique index "i" (see the first column in the matrix). Then, a set of indexes next to the particular mapping $A_{\{i\}}$ or $M_{\{i\}}$ specify, which UCM concepts are used to form a corresponding SBVR concept or rule. This feature and the logic behind it were not used in [18].

Let's take an example. There is a mapping $A_{1,2,3}$ at the intersection of UCM *Association* ($i = 3$) and SBVR *Verb Concept*. This means that in our UCD → BV&R approach each association (: *Association*) in a given UCD will be a subject for automatic extraction of a SBVR verb concept; in this mapping, additional indexes $i = 1$ and $i = 2$ indicate that each instance of this extraction also involves certain UCD actor (: *Actor*) and use case (: *UseCase*) connected to each other via that particular association. For example, having an association connecting 'Sales clerk': *Actor* and 'Create rental contract': *UseCase*, one can automatically extract SBVR verb concept 'sales_clerk *create* rental_contract' from this triplet.

Table 1. M2M mapping matrix for the UCD → BV&R approach

| Index | UML UCD (source) | SBVR BV&R (target) | | |
|---|---|---|---|---|
| | UCM Concept | General Concept | Verb Concept | Business Rule |
| 1 | Actor | $A_1$ | - | - |
| 2 | Use Case | $M_2$ | - | - |
| 3 | Association | - | $A_{1,2,3}$ | - |
| 4 | Constraint (on association) | $M_4$ | $M_4$ | $M_{1,2,3,4}$ |
| 5 | Include | - | $A_{1,2,3,5}$ | $A_{1,2,3,5}$ |
| 6 | Extend | - | $A_{1,2,3,6}$ | $A_{1,2,3,6}$ |
| 7 | Extension Point | $M_7$ | $M_7$ | $M_{1,2,3,7}$ |
| 8 | Generalization (between actors) | - | $A_{1,8}$ | - |
| 9 | Generalization (between use cases) | - | $A_{1,2,3,9}$ | - |
| 10 | Boundary | $A_{10}$ | - | - |
| 11 | Comment | $M_{11}$ | $M_{11}$ | $M_{11}$ |

It has to be mentioned that without using complex natural language processing (NLP) techniques, *automatic* extraction of SBVR concepts and rules (i.e. mappings denoted by "A" in the matrix) assumes the use of best modeling practices [1][2] by default, e.g. name of a use case should be composed of a present tense verb and a strong noun (subject). Let us rename the use case 'Create rental contract' from our example into 'Rental contract creation'. Then, the mapping $A_{1,2,3}$ would result in the extraction of text rumbling** 'sales clerk rental contract creation', which could not be automatically transformed into correct verb concept in SBVR business vocabulary. Nevertheless, semi-automatic BV&R extraction approach allows user to identify such cases as bad practices and refactor them before transforming into SBVR concepts; this may also result in the refactoring of UCD itself following best practices in use case modeling.

Another important remark is that the approach considers only the mappings (Table 1) and transformation rules (Section 4.2), which are relevant to the development of *SBVR business vocabularies* and *business rules*. Let us consider a generalization between use cases (UCD example in Fig. 2) "*use_case$_1$ generalizes use_case$_2$*". In our approach, such construction cannot/should not be straightforwardly transformed to a verb concept entry of *SBVR business vocabulary* for several reasons. First of all, this would result in an invalid SBVR concept construction: "customer makes car_booking generalizes customer makes walk_in_car_booking". Secondly, we state that generalizations between use cases is a subject to so called *model vocabularies*, rather than *business vocabularies*. In a *use case model vocabulary*, use case model-specific general concepts are introduced, e.g. 'use_case', 'actor'; in its turn, this then allows one to construct specific general concepts, verb concepts and business rules entries in a *model vocabulary*: the use cases "Make car booking" and "Make walk-in car booking" would be specified as general concepts 'make_car_booking' and 'make_walk_in_car_booking', both of which have a more general concept 'use_case'; after that, the generalization between these use cases would be specified as 'use_case 'make_car_booking' *generalizes* use_case 'make_walk_in_car_booking'' (or 'make_car_booking *generalizes* make_walk_in_car_booking'). This remark is also true for the extraction of *verb concepts* representing <<include>> and <<extend>> relationships between use cases. However, once again, this is not a subject for a *business vocabulary* and, therefore, not this paper as well.

### 4.2 The Algorithm

On the highest level of abstraction, the algorithm of UCD → BV&R approach is composed of three basic stages (Fig. 1):

– *Stage 1*: Extraction of text rumblings from the source use case diagram;
– *Stage 2*: Formation of SBVR business vocabularies and rules;
– *Stage 3*: Validation of the developed overall SBVR specification.

*Stage 1* deals with the automatic extraction of text rumblings from the source UCD. These rumblings are then used in the next stage to form SBVR concepts and rules. In *Stage 2*, the algorithm follows the basic principle of so called business rules "mantra" (followed from "Business Rules Manifesto" [16]), which states that business

---

** A term "*text* rumbling" represents an unstructured piece of textual information in a problem domain. This term is derived from a term "*business* rumbling", which was first introduced by B. von Halle in her paper "Back to Business Rule Basics" (in Database Programming & Design, 1994) and had more specific scope than a "*text* rumbling" (i.e. *business* domain).

rules are built on facts (verb concepts) and facts are built on terms (general concepts). Verb concepts and general concepts are the ones forming the basis of any BV; successively, one must have a BV in order to specify and manage business rules properly. In this paper, *Stage 3* is assumed as a manual process where the extracted overall SBVR specification is validated with business domain expert. The algorithm covering all of the aforementioned stages is described in more details in Table 2.

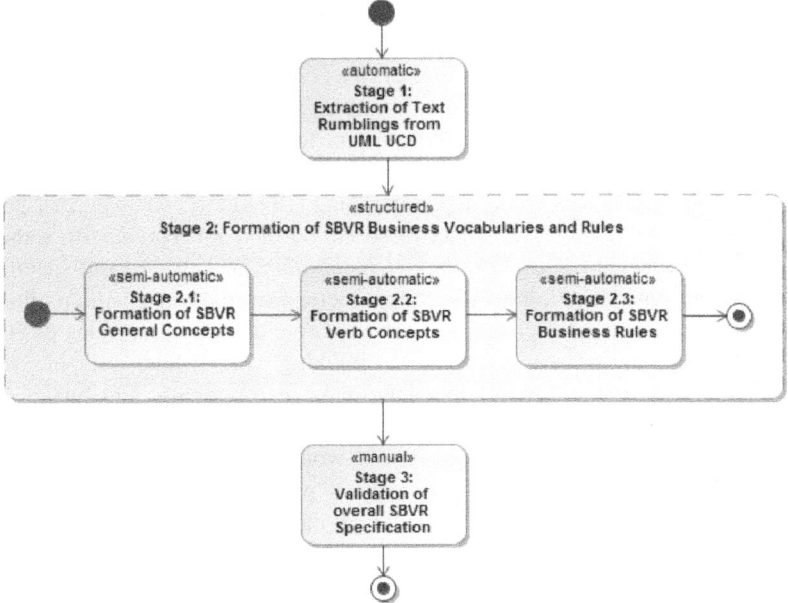

**Fig. 1.** Basic algorithm of the semi-automated UCD → BV&R approach

In Table 2, exemplary M2M transformations are specified by transformation rules written in pseudo code. General structure of these transformation rules are as follows:

- $transform(SourceModel, \{P_1, ..., P_j\}) \overset{A|M}{\longrightarrow} TargetConcept$,

where:
- *SourceModel* represents a source model for the transformation;
- $\{P_1, ..., P_j\}$ is a set of parameters representing concepts of the source model, which is required for the extraction of *TargetConcept* presenting a concept of a target model (i.e. SBVR BV&R); if any kind of relationship $P_i$ is passed as a parameter, then it is specified as $P_i(P_n, P_m)$, where $P_n$ and $P_m$ are two concepts bound to each other with $P_i$;
- In a certain transformation rule, transformation operator may be "$\overset{A}{\rightarrow}$" or "$\overset{M}{\rightarrow}$". Letter "A" denotes transformation, which is performed automatically by the transformation tool (such rule realizes certain "A" mappings from the mapping matrix (Table 1)). Letter "M" denotes transformation, which requires certain user interaction to realize certain "M" mapping from the mapping matrix (Table 1).

Also, a transformation may have a number of preconditions, which must hold true in order to enact that transformation. We denote a precondition with the tag <precond:> and the specify it's body in natural language text; see an example of precondition in Step 2.2.2.

**Table 2.** Description of the algorithm (Fig. 1)

| Stage | Description |
|---|---|
| 1 | All needed text rumblings are *automatically* identified and extracted from a given UML UCD. These rumblings are formed from the names of particular elements in the source UCD. Some meta-data about the rumblings is also extracted from the UCD; this includes types of the UCD elements the rumblings were extracted from, and also all kinds of relationships.<br>In M2M transformation wizard (top right GUI window in Fig. 2), a subset of the extracted text rumblings is presented in table column "Text rumblings"; this particular subset is composed of the rumblings for the extraction of SBVR verb concepts. |
| 2.1 | *Stages 2.1* and *2.2* are dedicated to the formation of SBVR BV. Stage 2.1 deals with the formation of the general concepts of that vocabulary.<br>Note that in the presented fragment of the implementation prototype, only the 2$^{nd}$ tab in M2M transformation wizard's GUI window is visible – this tab is dedicated to work with the formation of SBVR *verb concepts* (i.e. Stage 2.2), which means that the intermediate results of this particular stage (as well as Stage 2.3) are not directly visible in wizard's GUI window in Fig. 2. Though, the final result of the whole Stage 2 is visible in SBVR editors' window (low right corner in Fig. 2).<br>*Stage 2.1* is composed of three steps:<br>– *Step 2.1.1*: Selective presentation of a subset of text rumblings to a user. Only those rumblings are presented, which are specified as a source for the extraction of SBVR general concepts (see the column "General Concept" in the target model section of the mapping matrix (Table 1)).<br>– *Step 2.1.2*: Extraction of candidate general concepts (CGC) from the presented text rumblings. Term "candidate general concept" means that the extracted text expression has yet to be interpreted and validated by a dedicated SBVR editor (Step 2.1.3). Depending on the specified level of automation in the mapping matrix, extraction of CGC may be performed automatically or with user interaction. Further, we present few examples of automatic and manual transformation rules used for the extraction of CGC:<br>    – Automatic extraction of a CGC from an *actor* in a given UCD. The goal is to form a *general concept* representing an *actor*:<br>    $transform(UCD, actor: Actor) \xrightarrow{A} general\_concept: GeneralConcept$,<br>    e.g.: $transform(UCD\_1, \text{``Rental manager''}) \xrightarrow{A}$ 'rental_manager' |

| Stage | Description |
|---|---|
| | – Manual extraction of a CGC from a *use case* in a given UCD. The goal is to form one or more *general concepts* representing on or more noun objects in a *use case*: <br><br> *transform(UCD, use_case: UseCase)* $\xrightarrow{M}$ *general_concept: General-Concept*, <br><br> e.g.: *transform(UCD_1, "Create rental contract")* $\xrightarrow{M}$ 'rental_contract' <br><br> – **Step 2.1.3**: Formation of SBVR general concepts from the defined CGC. After user finished working with a set of CGC, they are exported to VeTIS editor [9], which SBVR syntax interpreter and validator. User overviews the formed SBVR general concepts; if errors are found, one returns to the Step 2.1.2 to make certain corrections and repeats Step 2.1.3 afterwards. At the end, a set of SBVR business vocabulary entries representing general concepts is formed. |
| 2.2 | Following the business rules "mantra", verb concepts may be formed after general concepts were extracted from a given UCD. <br><br> **Stage 2.2** is composed of three steps similar to Stage 2.1: <br><br> – **Step 2.2.1**: Selective presentation of a subset of text rumblings to a user. Rumblings are selected according to the markings in the column "Verb Concept" in the target model section of the mapping matrix (Table 1). <br><br> – **Step 2.2.2**: Extraction of candidate verb concepts (CVC) from the presented text rumblings. An important and useful feature in this step is the automatic identification of general concepts in the presented CVC (Fig. **2**.) – these general concepts are presented in predefined font style ('term'), which is used to represent SBVR general concepts. Few examples of automatic transformation rules used for the extraction of CVC: <br><br> – Automatic extraction of a CVC from a *generalization* relationship relating two *actors* in a given UCD. The goal is to form a *verb concept* representing a *generalization* between two *actors*: <br><br> *transform(UCD, actor$_1$: Actor, actor$_2$: Actor, generalization(actor$_1$, actor$_2$): Generalization)* $\xrightarrow{A}$ *verb_concept: VerbConcept*, <br> e.g.: *transform(UCD_1, "Rental manager", "Rental manager's assistan", generalization("Rental manager", "Rental manager's assistan"))* $\xrightarrow{A}$ 'rental_manager *generalizes* rental_manager_assistan'. <br><br> – Automatic extraction of a CVC from a triplet of an *actor*, *use case* and *association* relating the *actor* and the *use case* in a given UCD. The goal is to form a *verb concept* representing an *actor* performing a *use case*: <br><br> *transform(UCD, actor: Actor, use_case: UseCase, association (actor, use_case): Association)* $\xrightarrow{A}$ *verb_concept: VerbConcept*, |

| Stage | Description |
|---|---|
| | e.g.: *transform(UCD_1, "Rental manager", "Create rental contract", association("Rental manager", "Create rental contract"))* $\xrightarrow{A}$ 'ren<u>tal_manager</u> create <u>rental_contract</u>'.

– Automatic extraction of a CVC from a *generalization* relationship relating two *use cases* and an *association* relating an *actor* and a parent *use case* in a given UCD. The goal is to form of a *verb concept* representing an *specialized use case* and an *actor* inherited from a *more general use case*:

*transform(UCD, use_case$_1$: UseCase, use_case$_2$: UseCase, actor: Actor, generalization(use_case$_1$, use_case$_2$): Generalization, association (actor, use_case$_1$): Association)* $\xrightarrow{A}$ *verb_concept: VerbConcept*,

e.g.: *transform(UCD_1, "Make car booking", "Make walk-in car booking", "customer", generalization("Make car booking", "Make walk-in car booking"), association ("customer", "Make car booking"))* $\xrightarrow{A}$ '<u>customer</u> make <u>walk_in_car_booking</u>'.

– Automatic extraction of a CVC from an *include* relationship relating two *use cases* and an *association* relating an *actor* and a base *use case* in a given UCD. The goal is to form a *verb concept* representing an *included use case* and an *actor* inherited from a *base use case*:

<precond:> there is no *association* between the included *use_case$_2$* and any *actor* specified in a given UCD;

*transform(UCD, use_case$_1$: UseCase, use_case$_2$: UseCase, actor: Actor, include(use_case$_1$, use_case$_2$): Include, association (actor, use_case$_1$): Association)* $\xrightarrow{A}$ *verb_concept: VerbConcept*.

This transformation rule will not be enacted for the given UCD_1 (see the use case diagram (UCD_1) in Fig. 2) because the precondition is not satisfied for any *include* relationship in UCD_1.

– **Step 2.2.3**: Formation of SBVR verb concepts from the defined CVC. A set of CVC is exported to VeTIS editor [9]. If errors in verb concept entries are found, one returns to the Step 2.2.2 to make certain corrections and repeats Step 2.2.3 afterwards. After the Stage 2.2 is finished, the extraction of SBVR business vocabulary from a given UCD is completed. Next stage will use this BV to extract and form SBVR business rules. |
| 2.3 | SBVR business rules are built on the verb concepts from Stage 2.2. Compared to the formation of SBVR BV concepts, the formation of business rules is more complicated and time consuming; however, the steps in Stage 2.3 do not differ from Stages 2.1 and 2.2:

– **Step 2.3.1**: Selective presentation of a subset of text rumblings to a user. Rumblings are selected according to the markings in the column "Business Rule" in the target model section of the mapping matrix (Table 1). |

| Stage | Description |
|---|---|
| | – **Step 2.2.2**: Extraction of candidate business rules (CBR) from the presented text rumblings. Previously specified verb concepts are automatically identified and presented in predefined font style ('<u>term</u>', '*verb*') in the extracted CBR expressions. Few examples of transformation rules used for the extraction of CBR: |
| | – Automatic extraction of a CBR from an *include* relationship relating two *use cases* in a given UCD. The goal is to form a *business rule* representing an obligation for an *actor* to perform an *included use case*: *transform(UCD, use_case₁: UseCase, use_case₂: UseCase, include(use_case₁, use_case₂): Include)* $\xrightarrow{A}$ *business_rule: BusinessRule*, e.g.: *transform(UCD_1, "Create rental contract", "Manage rental insurance", include("Create rental contract", "Manage rental insurance"))* $\xrightarrow{A}$ 'It is obligatory that <u>rental_manager_assistant</u> manages <u>rental_contract</u> if <u>rental_manager</u> creates <u>rental_contract</u>'. |
| | – User interacted extraction of CBR, which is initiated by existing *association condition* in a given UCD. The goal is to form a *business rule* representing a permission for an *actor* to perform a *use case* if certain condition is satisfied: *transform(UCD, actor: Actor, use_case: UseCase, association(actor, use_case): Association, association_condition: Constraint)* $\xrightarrow{M}$ *business_rule: BusinessRule*, e.g.: *transform(UCD_1, "Rental manager's assistant", "Create rental contract", association("Rental manager's assistant", "Create rental contract"), "sales manager is absent")* $\xrightarrow{M}$ 'It is permitted that <u>rental_manager_assistant</u> creates <u>rental_contract</u> if <u>rental_manager is_absent</u>'. |
| | Note that the latter transformation rule illustrates an example where the user interaction is needed in order to extract a valid business rule. This is due to the fact that there is no standard template (best practice) available to specify association conditions in UCD; therefore, the resulting CBR should be manually validated (and corrected if needed) by a user. |
| | – **Step 2.3.3**: Formation of SBVR business rules from the defined CBR. This step is similar to the corresponding steps in previous stages. |
| 3 | Final validation of SBVR BV&R with a business domain expert is done to ensure semantic validity of the specification and its consistency with the original source model. It may invoke multiple iterations with the previous stages; this may also result in refactoring of the original source model. |

## 4.3 The Implementation

On the implementation level, the proposed approach may be viewed as an interaction of three dedicated systems:

Approach for Semi-automatic Extraction of Business Vocabularies and Rules    193

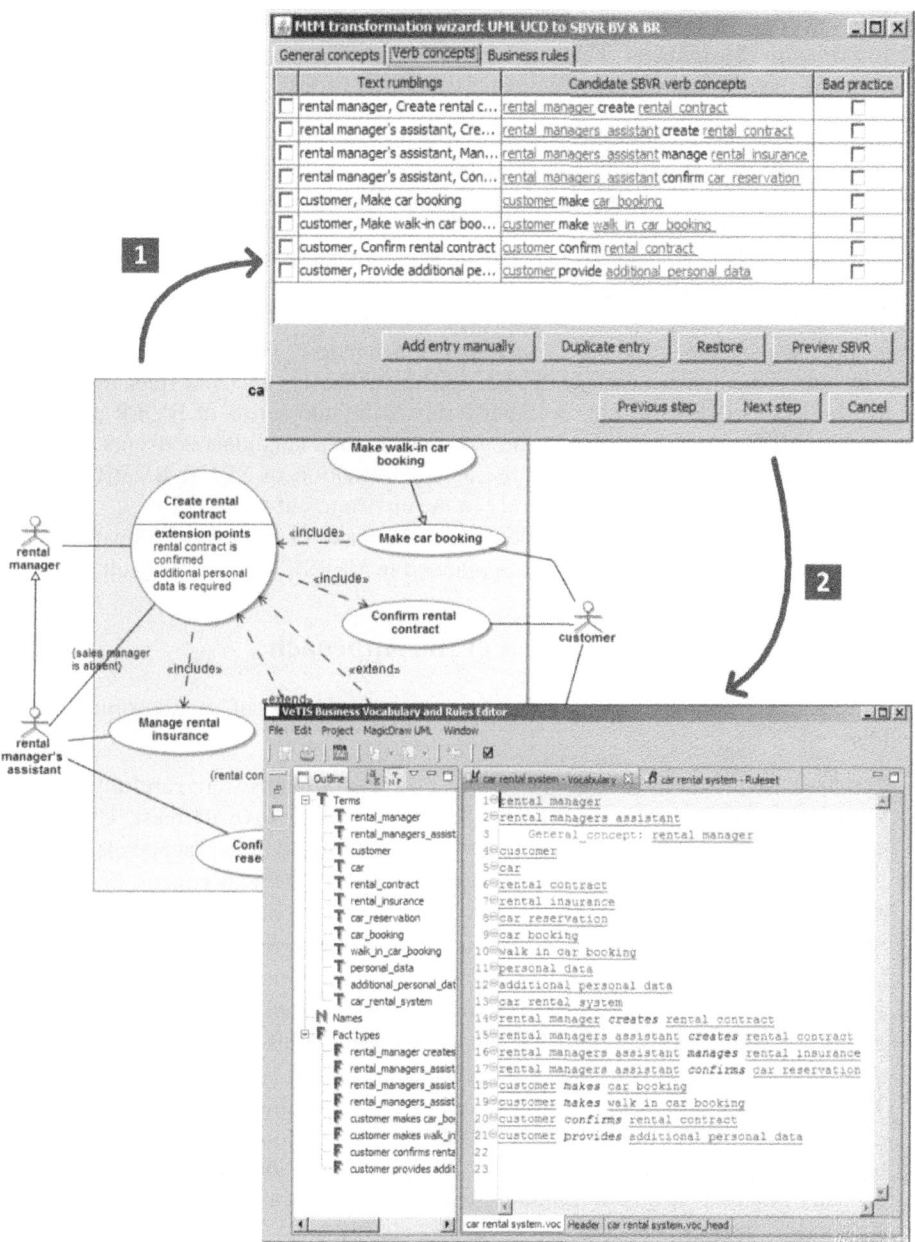

**Fig. 2.** Interaction of MagicDraw, UCD → BV&R tool and VeTIS tool

- CASE tool *MagicDraw*. The tool supports the newest official versions of UML, BPMN and other modeling standards; however, it still lacks support for SBVR. *MagicDraw* has *Open Java API*, which enables the development of plug-ins.
- Hard-coded UCD → BV&R tool working as a plug-in of the *MagicDraw*.

- SBVR editor VeTIS [9]. This is yet another tool developed by the Department of Information Systems at Kaunas University of Technology, Lithuania. VeTIS supports the development and syntactic validation of SBVR BV&R.

At this point, the main engineering objective of the research is to extend the functionality of *MagicDraw* tool by adding new UCD → BV&R extraction feature.

In Fig. 2, Label 1 shows that text rumblings are being extracted from a given UCD, which was modelled using *MagicDraw* (this implements Stage 1 of the previously introduced algorithm). The main work with the extracted text rumblings is performed using a hard-coded BV&R extraction tool; this tool implements the whole wizard-assisted extraction process of the approach. According to business rules "mantra", user starts with general concepts in the first tab window; then the next tab window provides functionality to work with verb concepts; finally, business rules are formed in the third tab window. In Fig. 2, Label 2 shows the interaction of BV&R extraction tool with VeTIS editor; this interaction is invoked when candidate concepts and rules are being transferred to VeTIS, i.e. the actual formation of SBVR BV&R is being performed. Stage 3 is performed entirely in the environment of VeTIS tool.

As of yet, there is no implementation to make a backward transformation (i.e. BV&R → UCD). This, however, is planned in VEPSEM project as well.

## 5  Experimental Evaluation of the Approach

To evaluate the approach, certain experiment was carried out. It was organized as follows:

- Ten use case diagrams describing various problem domains were carefully selected from the internal database of BSc/MSc thesis works. Five of these UCD were sound diagrams and other five contained certain bad modeling practice elements.
- A group of business/system analysts was selected for the experiment. Each expert had good practical experience working with UML; however, their knowledge in SBVR varied from "basic" to "expert". All experts were instructed about the UCD → BV&R approach.
- Each expert was provided with five (out of ten) randomly selected use case diagrams and asked to extract SBVR business vocabularies and business rules from those diagrams using the UCD → BV&R approach.
- The results as well as the overall performance were evaluated against certain criterions: rates of successful extraction of SBVR general concepts, verb concepts and business rules; number of errors caused by the experts; identified number of errors and bad practices in given diagrams while performing the task; time required to accomplish the task; experts' feedback after applying the approach.

After analyzing the results, certain conclusions were made:

- Experts did not have any difficulties performing automatic or manual extraction of general concepts and verb concepts from the sound use case diagrams. Diagrams with inconsistent naming schemes provided higher rate of errors in the results of low-experienced SBVR users; for experienced SBVR users, this was not an issue – they dealt with naming inconsistencies fluently.

- Low-experienced SBVR users also struggled extracting valid candidate business rules from text rumblings. Though, the syntactic errors they committed were afterwards corrected using VeTIS editor (again, VeTIS supports SBVR syntax validation feature). We also identified few cases of semantically incorrect business rules, which were correct SBVR syntax-wise – such cases are hard to detect and validate without good knowledge of the problem domain itself.
- As expected, during the BV&R extraction process, users detected certain invalid uses of UML syntax and bad practices in the given diagrams. This proved our presumptions that the quality of use case diagrams can also be improved during the extraction process or after it.
- Comparison of task execution times showed that it could take almost twice as much time to complete the task with unsound source diagrams compared to the time needed to extract BV&R from the sound ones. This is true for both experienced and low-experienced SBVR users. Of course, in both cases experienced users showed shorter execution times compared to low-experienced users.

Overall feedback from the experiment participants was positive. A guided (wizard-based) BV&R extraction process was distinguished as a main positive feature of the approach. However, for actual practical application, the approach should provide even higher level of automation (e.g., by utilizing benefits of natural language processing technologies) and more functionally enriched tool.

## 6 Conclusions

Information systems developers do agree that to have a well-formed business vocabulary in any ISD project is nearly a must. One of the newest OMG's standards in the area of business modeling (which is a part of ISD) is "Semantics of Business Vocabulary and Business Rules" [15] providing the infrastructure for such vocabularies as well as more complex business statements – business rules. However, in actual IS developments, this practice is still not a common case scenario. We believe that the main reasons for this are rather complex development and management of such specifications ("will it actually pay off?"), the lack of CASE tools' support, and a relatively short time-in-market of the SBVR standard itself.

We argue that BV&R specifications should not be necessarily developed from scratch – one can extract them from other existing business models (e.g. business process models as shown in [18]) or IS models in the early stages of ISD. In this paper, an approach for semi-automatic extraction of SBVR business vocabularies and rules from UML use case diagrams is presented. The extracted BV&R specifications can then be elaborated and used further on in ISD life cycle and beyond. Our research confirmed that SBVR BV&R can indeed be extracted from UML use case diagrams.

Certain good modeling practices should be involved in order to increase the effectiveness of the approach. At the same time, one must acknowledge that the need to follow certain modeling rules may act as an off-factor for some modelers. This could be partially solved by introducing these modeling practices directly into CASE tools in a form of non-critical recommendations or by other appropriate means.

Overall, we conclude that a full implementation of UML UCD → SBVR BV&R approach would provide certain benefits, such as: faster development of well-structured, formalized business vocabularies and rules; increased quality and completeness of the business model, which in its turn could affect the quality of further developments, e.g. IS design. Further research could result in the development and implementation of UCD↔BV&R two-way transformation (and synchronization) approaches. The development of fully-automatic transformation approach is also considered in our near future plans.

## References

1. Adolph, S., Bramble, P., Cockburn, A., Pols, A.: Patterns for effective use cases. Addison-Wesley (2002)
2. Ambler, S.W.: The Elements of UML 2.0 Style. Cambridge University Press (2005)
3. ATLAS group, LINA & INRIA: ATL (Atlas Transformation Language) User Guide (2014), http://wiki.eclipse.org/ATL/User_Guide
4. Cabot, J., Pau, R., Raventós, R.: From UML/OCL to SBVR specifications: A challenging transformation. Information Systems 35, 417–440 (2010)
5. Deeptimahanti, D.K., Sanyal, R.: Semi-automatic generation of UML models from natural language requirements. In: Proc. of the 4th India Conference on Software Engineering (ISEC 2011), pp. 165–174 (2011)
6. Georgiades, M.G., Andreou, A.S.: Formalizing and Automating Use Case Model Development. The Open Software Engineering Journal 6, 21–40 (2012)
7. Jouault, F., Kurtev, I.: On the interoperability of model-to-model transformation languages. Science of Computer Programming 68, 114–137 (2007)
8. Kärkkäinen, T., Nurminen, M., Suominen, P., Pieniluoma, T., Liukko, I.: UCOT: semiautomatic generation of conceptual models from use case descriptions. In: Proc. of the Int. Conf. on Software Engineering (IASTED), pp. 171–177 (2008)
9. Nemuraite, L., Skersys, T., Sukys, A., Sinkevicius, E., Ablonskis, L.: VETIS tool for editing and transforming SBVR business vocabularies and business rules into UML&OCL models. In: Proc. of the Int. Conf. on Information and Software Technologies (IT 2010), Kaunas, Lithuania, pp. 377–384 (2010)
10. Njonko, P.B.F., El Abed, W.: From natural language business requirements to executable models via SBVR. In: International Conference on Systems and Informatics (ICSAI), pp. 2453–2457 (2012)
11. NL2OCL Project. NL2OCLviaSBVR – A Natural Language to OCL Transformation via SBVR, http://www.cs.bham.ac.uk/~bxb/NL2OCLviaSBVR/NL2OCLviaSBVR.html
12. OMG. Business Process Model and Notation (BPMN) v.2.0. OMG Doc. No.: formal/2011-01-03 (2011)
13. OMG. Meta Object Facility 2.0 Query/View/Transformation Specification v1.1 (2011)
14. OMG. Model-Driven Architecture (MDA) v.2.0 (2003)
15. OMG. Semantics of Business Vocabulary and Business Rules (SBVR) v.1.1 (June 2012)
16. OMG. The Business Rules Manifesto (2003), http://businessrulesgroup.org/brmanifesto.htm
17. OMG. Unified Modeling Language (UML), Superstructure v2.4.1 (2011)
18. Skersys, T., Butleris, R., Kapocius, K., Vileiniskis, T.: An Approach for Extracting Business Vocabularies from Business Process Models. Information Technology and Control 42, 178–190 (2013)
19. Thakore, D., Upadhyay, A.R.: Development of Use Case Model from Software Requirement using in-between SBVR format at Analysis Phase. International Journal on Advanced Computer Theory and Engineering (IJACTE) 2, 86–92 (2013)

# Author Index

Aerts, Walter    76
Araki, Akiyoshi    31
Aveiro, David    46, 105

Blair, Gordon    151
Butleris, Rimantas    182

Cruz, Estrela Ferreira    167

Danenas, Paulius    182
Décosse, Céline    16
de Jong, Joop    91
de Vries, Marne    1
Drobek, Marc    151

Figueira, Carlos    46

Gerber, Aurona    1
Gilani, Wasif    151

Hunka, Frantisek    61
Huysmans, Philip    76

Iijima, Junichi    31, 136

Kalidindi, Vishnupriya    120

Liu, Yang    136

Machado, Ricardo J.    167
Molka, Thomas    151
Molnar, Wolfgang A.    16

Pombinho, João    105
Proper, Henderik A.    16

Rashid, Awais    151
Redlich, David    151

Santos, Maribel Yasmina    167
Skersys, Tomas    182

Tribolet, José    105

van der Merwe, Alta    1
Vanhoof, Els    76
Verelst, Jan    76

Wan, Yun    120

Zacek, Jaroslav    61

The manufacturer's authorised representative in the EU is Springer Nature Customer Service Centre GmbH, Europaplatz 3, 69115 Heidelberg, Germany. If you have any concerns regarding our products, please contact ProductSafety@springernature.com

Printed and bound by CPI Group (UK) Ltd, Croydon, CR0 4YY
23/03/2026
02076673-0018